THE INTUITIVE BEEKEEPER

THE INTUITIVE BEEKEEPER

Beyond Master Beekeeping

JONATHAN ADAM HARGUS

Charleston, SC
www.PalmettoPublishing.com

The Intuitive Beekeeper

Copyright © 2023 by Jonathan Adam Hargus

All rights reserved

No portion of this book may be reproduced, stored in a retrieval system, or transmitted in any form by any means–electronic, mechanical, photocopy, recording, or other–except for brief quotations in printed reviews, without prior permission of the author.

First Edition

Paperback ISBN: 979-8-8229-2562-5

DEDICATION

To my wife Haleigh, who not only supported me like a superstar teammate in the writing of this book but also taught me a new way of beekeeping without knowing anything about it. Thank you love…

Table of Contents

Foreword ix
Introduction 1
Chapter 1 Assessing a Hive Before Inspection. 5
Chapter 2 Assessing a Hive During Inspection 16
Chapter 3 What to Do with Your Assessment. 28
Chapter 4 Anticipating Your Bees' Needs. 41
Chapter 5 Working with the flow 50
Chapter 6 Working with the Bees 57
Chapter 7 Queen-Keeping. 64
Chapter 8 The 7 Methods of Minimal Disturbance. 75
Chapter 9 Mindfulness & Intention Allows for Intuition. . . . 86
References 97
Glossary of Beekeeping terms & Lingo 99

Foreword

Jonathan and I connected through my blog wayward-bee.com as we had a similar approach to our beekeeping, in that we were both practicing beekeeping methodologies that supported the bees to be productive through their natural behavior rather than the more traditional way of overriding these instincts in order to gain a large honey crop, often to the detriment of the bees themselves.

We also discussed natural forage, the nectar and pollen producing herbs, trees, weeds and vegetables that the bees rely on for food. We agreed on the importance of providing a wealth of variety and diversity, and how this encourages a greater abundance for other species as well as our honey bees.

Given the challenges with species loss and rapid changes in the biosphere in our modern world, this consideration for a thriving natural environment is of great benefit to all living beings, ourselves included. Bees will gather nectar some distance from the hive, and just as with us humans, a poor diet and toxins will affect the health and vitality of a bee colony. By helping our bees become fit and productive, we create advantageous conditions for other bees, other insects, birds, mammals, reptiles and of course the plants and microorganisms on which all of us ultimately depend.

Living on either side of the Atlantic, our experiences with beekeeping were interestingly comparable and different (*here in England, I don't have to deal with bears!*) so we found a useful sounding by sharing our findings. I have no formal beekeeping qualifications, but use my wealth of natural history knowledge to guide me; Jonathan has worked with large beekeeping operations and knows what is involved at the more industrial end of the spectrum. For me, as someone who feels so strongly about the ability of bees and honey to engage the public and encourage a harmonious relationship with our environment, this transition of Jonathan's from commercial to compassionate is admirable, and I have enjoyed seeing how expertly and confidently he handles his bees, and how his understanding of how a bee colony functions across the year and relates to its immediate surroundings is both informative and inspirational.

Beekeeping has a reputation for being something only tough people can do, with images of swarming insects, stings and clouds of smoke, as well as jars of golden honey stacked high. Actually, beekeeping allows us to commune with these hardworking and hyper-organized beings in a way that we just can't with any other pet or hobby, and Jonathan's instructions to draw on that forgotten connection that each of us have with nature —our intuition —will have a wider reach beyond the hive. Tapping in to an ability to "read" the weather, the temperature, the bees' mood, the trees and flowers in front of us is something we are all innately capable of yet rarely bother to consider. This message is much more palatable with the reward of delicious and nutritious honey at the end which by ensures the bees have plentiful food and favorable conditions; the culmination of the bees' hard work under our care.

I wish Jonathan every success with this book, and look forward to a new and innovative paradigm in the world of beekeeping, working with nature, with the bees, and discovering an affinity with these legendary insects that will serve us, and our planet, into the future.

—Jennifer Moore, *beekeeper, author, artist, instructor*

Introduction
—The Intuitive Beekeeper—

In my experience, intuition is something that we all have within us but that most people ignore. It's like an untapped insight into greater truth that has always been there and is always available to us as 'enlightened' creatures but that we are almost trained by modern society to ignore and therefore unknowingly stifle.

There seems to be a lack of and therefore a greater need for balance between the mind (*logic*) and the heart (*feeling*). If you are like me, you mostly use your mind to rationalize things and solve your everyday challenges, searching for solutions as you navigate through life without much consideration for your feelings which just get in the way.

On the flip side, those who mostly use their heart to guide them tend to lack the discipline of the mind to accurately steer their emotions and likewise end up where they don't want to be.

In my opinion, we need a balance. It is my goal in this book for you to see how this balance comes into fruition, helping to create a much more rich experience in the world of beekeeping where confidence overrides ignorance.

Let's look at our emotions for a minute. Our emotions are good, they are our compass and guidance system. This compass tells us where we are and where we are going and how we feel about everything along the way. However, it doesn't necessarily create the circumstances for getting where we need to go and where we want to go safely.

That's where our mind comes into play. Our logic and reasoning should be the safeguard that our hearts need in order to get us where we want to go. Unfortunately most of us allow our mind to be in control, to steer the ship if you will. The downside of letting our mind sit in the captain's seat is that our mind likes to play it safe and we miss tons of potential opportunities in life for advancement on many levels.

In actuality, our heart should be the captain of our vessel and our mind at the helm. Together they create a beautiful balance of purpose and intention, giving us the ability to create the results that we desire and help us travel to where we truly wish to be and it's almost always more than we imagined along the way.

When that balance harmonizes within us as beekeepers, and with anything else in our lives for that matter, we have discovered the secret for tapping into something that takes us far beyond being beginners, far beyond hobby beekeeping, and yes even beyond master beekeeping; we become *Intuitive* Beekeepers.

In this book, I am taking on the role as your guide for tapping into that lost art we call intuition so that you will realize the full potential of your own abilities and take them as far as you want to go. You will never approach a beehive the same way again.

It is therefore my personal goal which I have set forward to communicate through this book, that I state what we are to accomplish:

The need for a better way of beekeeping, ideally methods and practices involving Biodynamics & Sustainability.

The importance to understand the relationship of the honey bee to its floral environment and the unique role that plants have; specifically with honey bees.

And a need to slow down and become intentional in our beekeeping operations, prioritizing the threefold relationship between: the honey bee colony as a collective of individuals creating a single soul-entity, the plant kingdom and a heart-centered beekeeper who profits as a *result* of the cohesion of these three elements working together as one rather than *profit* being the primary objective.

You will find this book can be redundant and repetitive at times. The intended redundancy is for two reasons: The first reason is to drive home the principles in this book to the reader for a better understanding of beekeeping practices overall. The second reason is because any system that is designed with redundancies cannot fail because when one part fails, or in this case is misunderstood, there is always another part from which the system as a whole can still be effective and therefore realized.

"Read on!"
—Marcel the Shell

CHAPTER 1

ASSESSING A HIVE BEFORE INSPECTION

Evaluating a beehive begins before removing the hive cover. Ideally, a beekeeper will learn to know what to expect concerning the condition on the inside of the colony depending on the outside conditions, i.e. the time of year in his or her climactic location.

Granted, much of this comes with experience but if you do not properly employ your experience then you will find yourself always failing and not understanding why.

Intuition is a descendant of wisdom and wisdom is a product of knowledge coupled with experience. If wisdom is not actually applied from lessons learned then your knowledge and experience will always lead you to the same dead ends such as:

- Your hives swarming unexplainably.
- Your hives do not make it through the winter (*it's not necessarily because of the cold temperatures*).
- Your hives seem to always have queen issues.
- Or perhaps your bees fail to make a crop of honey.

These are just a few basic examples of what happens when your hives fail and when they do fail, you feel like a failure as a beekeeper.

The majority of beginners give up during or after their first year in beekeeping.

I'm here to encourage you that by applying wisdom based on your knowledge and experience, as little as it may be, it will help you fail your way to success in beekeeping because with these elements in place, you will develop the intuition needed to keep bees with greater ease by learning what works and what does not.

By learning from our mistakes and from the bees themselves as our tiny but silent instructors, being a better beekeeper will happen much faster than you realize because you will begin to see beekeeping from a completely different angle.

We can only learn so much from beekeeping books. I have my own collection of beekeeping books and I can tell you that it is very small because I am very particular about the type of knowledge I consume. What's needed beyond book learning is experiential learning; learning by doing.

At this point I'm going to give you some elements that help us in assessing our hives, some things that you may have never considered when it comes to beekeeping. Let's begin with the primary element that all beekeepers use and are affected by: the Seasons.

TIME OF THE YEAR

It's no secret that we have four seasons on this earth and there are few beekeeping books that break beekeeping down into its seasonal tasks. Applying seasonal beekeeping to our practices can be a serious challenge however. Do you know when spring truly begins in your area? It's not necessarily during the Vernal Equinox according to your wall calendar.

In my area of the north Georgia mountains, spring begins for my bees in late February. It's definitely still cold outside and considered wintertime but this is when maple trees start to bloom. On the warmer days when the high temperature is in the mid to upper 50's, my bees bust out of the hive like fighter jets taking to the sky.

They return just as quickly and clumsily with legs loaded with pollen collected from the maple trees. With this influx of pollen coming into the hive, the queen knows that she can increase her egg-laying rate because the colony has the resources to support colony increase. The days are getting longer and it is at this time that the population of our hives will exponentially increase during the following weeks and into the summer.

So the lesson to learn here is that when I see maple trees blooming and the days are getting warmer, at least in the mid 50's when it's warm enough for bees to fly, I know that my hives are taking every advantage of that time to gather much needed resources after a few months of winter. I don't need to look into my hive to verify that the queen is laying more eggs, I already know that external circumstances are ideal for my bees and what they do as a result of those circumstances. This is realizing the season and what it offers and also how my bees respond to those circumstances.

Keep this in mind however, spring to me and my bees and in my climactic location may not be the same as yours. If you are a Florida beekeeper for example, you could have pollen sources such as maple or willow blooming as early as sometime between Thanksgiving and Christmas!

Perhaps you are a beekeeper in Maine or Alaska. Your spring is going to begin much later than it does in my area. No matter where you are, spring is spring. Whether it's later or earlier, you can count on the bees doing the same thing they do every spring: gathering pollen and foraging for nectar which means they are increasing their egg-laying rate and thereby their population.

One of the most wonderful things about bees is that no matter where you are or what season you are in, they continue to be instinctually consistent according to the seasonal and geographical elements involved. Knowing this and remembering this will automatically give you a wisdom point if you apply it which leads us to the first key of successful beekeeping:

- **Key #1: Educate and learn your area's weather, climate, seasons, and honey bee forage and when it's available… and learn it well.**

So how does knowing external hive conditions such as bloom time, and internal hive conditions such as an increased egg-laying rate help? It tells us as beekeepers what to expect next.

Here is the second secret to successful beekeeping, so read carefully:

- **Key #2: The successful beekeeper is always thinking one season ahead.**

What does this mean? Basically it means that you must come to learn and anticipate the bees' needs according to the seasons and then to prepare for them ahead of time.

Beekeeping is not a procrastinator's game. At the time of writing this chapter, I am in middle to late summer. My thoughts and plans currently involve finishing up my summer tasks in a way that prepares my hives for fall and ultimately winter.

I'm thinking one season ahead so that I do not get behind because waiting until it's time to take action means it's too late. I have been too late before and guess what? I always pay for it. I have lost hives, I have lost out on larger honey crops and I have lost out of being more successful. It does not feel good nor does it do my confidence any favors.

How to Assess a Hive Before Inspection

So how does intuition play a part in assessing a beehive before inspection after factoring in the climactic and seasonal conditions?

Ideally and under the best of circumstances, these elements will tell the intuitive beekeeper what a healthy hive should look like at a specific time of the year.

A very basic example of this is that during the summer, when hive populations are at their highest, you should expect to see a hive billowing with bees upon opening the top cover. For these conditions to be met it means the following simple things:

- They have a healthy laying, quality queen.
- They have plenty of food stores (*pollen & nectar*) and resources are coming in.
- And they more than likely have lots of healthy brood.
- Which in turn means they have *manageable* levels of in-hive pests such as mites or hive beetle for example.

So let's say that we approached a hive during the summertime for inspection and we saw very few bees coming and going from the hive entrance with a small population.

Right away we should know that something is wrong because it does not look like a hive should look like at this time of the year. We do not know what the problem is yet but our expectations of what a hive should look like in summer have not been met.

The first pre-inspection clue of this hive's condition was that there were very few bees coming and going from the hive entrance.

In most areas, summertime means that there are summer flowers blooming everywhere. So the fact that there are few foragers in our example gives us our first indication that something is amiss. At this point, to continue our example, and knowing that there is plenty of bloom available for forage, our instincts should tell us that this hive will most likely have little to no food stores.

There are exceptions to this rule when there is a high Varroa Mite count but for the sake of our goal, we will continue on and come back to that shortly.

Upon opening the hive cover and seeing bees between two to four frames instead of all ten confirms our suspicions of a low population count. And looking down between the frames will tell us how many frames of honey they have for food or the lack thereof.

When a hive has a low population, they have less foragers and therefore less resources to gather more resources. Our observance of the hive entrance having little activity has confirmed our suspicions of the internal condition of the colony.

Now that our pre-inspection assessment of the hive's condition has been confirmed by removing the hive cover and inspecting, it's time to figure out *WHY* the hive has a low population.

Since there could be multiple explanations, let's look at some of the possibilities:

- The colony could be queen-less. Why? Who knows? Perhaps the queen was too old and they tried to raise a new queen but failed to do so. Maybe you accidentally squished her during the last hive inspection. There's not always a way to know how a hive becomes queen-less and that's okay. What is important is how to move forward from here which we cover in *Chapter 3: What to do with your assessment.*
- The hive could be droney. What does that mean? It means one of two possibilities: 1) The queen failed to adequately mate and is now '*shooting blanks*'. In other words she is all out of the ability to lay fertilized eggs and is now only laying *un*fertilized ones which all develop into drones. Or 2) Due to unknown circumstances, the colony now has a laying worker bee. Yes, workers can and do lay eggs in queen-less situations however they are all unfertilized eggs, again resulting in drone bees which do not contribute to gathering colony resources.
- Let's look at one more possibility just to broaden our example; the hive could have a high Varroa mite count and also be highly hygienic, removing all affected brood from the developing brood nest and discarding the affected brood from the colony, thereby resulting in not only an apparent poor brood pattern but a hive that cannot grow because of increased exposure to Varroa-spread viruses.

There are other possibilities of course but these are among the most common. We will talk more about assessing hives during an inspection in Chapter 2.

One of the most challenging things in beekeeping is that there is so much to learn. I can remember back when I first had beehives and thinking how cool it was to have bees and yet at the same time, I didn't know what to do with them much less when to do it.

I had absolutely no concept of the seasons in regards to beekeeping and therefore did not know how to prepare. When I looked into a hive it was purely out of curiosity and the feeling that I should be doing something. Needless to say, those hive did not last very long.

What changed my beekeeping game was when I started paying attention to the seasons and observing my bees, plus I had a mentor for fifteen years before ever getting my own bees again.

Earlier I used maple bloom as an example of when spring begins for my bees. I now know when to expect maple to bloom every year. On top of that, I know how my hives are going to benefit from it and what to expect to see when it's time for the first spring hive inspection.

The essence of assessing a hive before inspection is learning to hone your own observation skills of your particular climactic region, your seasons and what blooms during each season and when those things start to bloom and for how long, and what an ideal hive looks like at each point throughout.

So how do you learn what blooms in your area? And how do you know if it's a plant that honey bees forage from or not? Honey bees are picky when it comes to foraging. They generally return to sources of nectar that offer the most bang for their buck, avoiding the minor sources of nectar until the major ones no longer offer them anything else which is why they are not the best pollinator for backyard gardens like other native, local pollinators are.

Increase Your Skills of Observation

There are three things you can do aside from in-hive inspections that will help you level up in your seasonal observation skills when approaching a hive:

- Join a local beekeeping club. I joined a club in my area purely to learn what blooms and when it blooms and for how long. After I learned that I started to learn what these plant sources looked like. The idea is to find other beekeepers **and learn from them.** You don't have to reinvent the brood box so to speak.
- One of the few books in my beekeeping collection is a book called, '*American Honey Plants*' by Frank C. Pellett. You will see me refer to it often. I recommend this book to every beekeeper, especially if you are ignorant of plants. It goes over each state in the U.S. and lists the major and minor sources of forage for honey bees.
- The third thing is very simple: go to your apiary and just sit and watch the hive entrance for some time. Watch the color of pollen the bees are bringing in and the rate at which they're doing it. Listen to the sound and frequency of their buzz; learn it. Smell the scent in the air of your apiary of curing honey or anything else in the environment. A wonderful book for guiding you through something like this is called *At the Hive Entrance* by Storch.

I'm going to give you an example from Pellett's book, *American Honey Plants* about my area in the north Georgia mountains to give you an idea of what you can expect.

"The willow, wild cherry, hawthorn, blackberries, raspberries, locust, holly and tulip tree (*Liriodendron tulipifera*) bloom in April. The two latter are most valuable for honey. The holly blooms for for about two weeks—the height of its flowering is about the first week in May. The tulip tree blooms for three weeks. This is the poplar tree of the south."

Take your beekeeping game to the next level by first teaching yourself what resources bees need to thrive and what plants in your area provide those resources. Next, teach yourself when those plants provide nectar and pollen and for how long. In my experience, nothing lasts more than four to six weeks when it comes to bloom time. There's always something new blooming whether it's from a major or minor nectar source.

In the next chapter, we are going to learn how to assess a hive during an inspection and how intuition plays a role in the inspection process. Remember, every time you learn something through experience you gain knowledge and every time you apply that experiential-based knowledge, you are applying wisdom.

Your wisdom is what is going to ripen over time to give you the intuition you need to be a successful beekeeper which leads us to the third key to successful beekeeping:

- **<u>Key #3</u>: The successful beekeeper learns from his or her mistakes.**

Eventually, you will come to know what to expect and know how to fix things when they aren't up to spec. One of the most common questions that I get from my students is, "*how many times should I be inspecting my hives?"* Which is a natural question when it comes to gaining experience, especially when you don't know what to do or when to do it.

We are going to answer this question in *Chapter 8: The 7 Methods of Minimal Disturbance* later in the book and it's probably not what you are expecting.

CHAPTER 2

ASSESSING A HIVE DURING INSPECTION

As you learned in the first chapter, assessing a hive begins before an in-hive inspection. When it comes to actually opening a beehive, there are really only two things this accomplishes:

- First, to confirm that the colony's condition is as expected according to the time of year it is. This will help to give you something to gauge your knowledge and experience against.
- And secondly, to consider what actions you as the beekeeper must take in order to help the bees to maintain the most ideal conditions and in-hive environment for them to thrive at this time of year.

The first goal simply helps to hone your beekeeperness. When you go into a hive knowing what to expect, the conditions you meet inside will either confirm your instincts and thereby strengthen them or they will raise a flag, drawing your attention to the fact that something isn't right, and call for further inspection. Either way you are learning.

Each season calls for certain tasks from the beekeeper. These tasks are your goals and reasons for going into a hive for inspection.

Never go into a hive out of curiosity, this is a high risk reason which we will go into greater depth about in *Chapter 8: The 7 Methods of Minimal Disturbance*.

The tasks of each season is different and the season will dictate the actions involved which also depend on how frequently you inspect a hive. In a minute we're going to get into some specific seasonal task examples but there's something you need to know first. No matter what season it is, you must always look for the following elements during all hive inspections throughout the year:

- Determine whether the colony has enough food stores. This means that you need to see if they have frames of honey or open nectar available. There should always be capped honey in a hive which is considered *'food for later.'* If a colony only has open nectar, it's good that they're bringing that into the hive but it's a bit concerning that they haven't been able to gather surplus which is something that honey bees are known for. You should also check that there is pollen packed in the cells around the brood nest to feed developing bees with.
- You should also look for evidence of a good laying queen. Notice that I did not say that you should look for and find the queen every time you do a hive inspection. Listen, the longer you are in a hive, the more you put the queen's mortality at risk as you raise, lower and manipulate frames with your cumbersome hive tool. There's only one queen in a hive and you don't want to put her at risk more than you need to. So by looking for *evidence* of the queen you are keeping her safer and practicing better beekeeping methods. Evidence of

a good laying queen looks like this: 1) Eggs are present in the cells. 2) A good laying pattern on the frame. In other words there's lots of eggs and they are surrounded by slightly older larvae which are also surrounded by older larvae yet, and finally that is or will soon be surrounded by capped brood cells.

- One last thing to remember for every hive inspection: Always leave the colony capable of raising a new queen. To do this, you simply need to make sure there are eggs from which the hive can raise a new queen from. The reason for this is because we could unknowingly kill the queen at any point during an inspection.

Now we're going to go through some examples of seasonal tasks starting with spring. We won't cover all beekeeping tasks because there's a lot but I will cover some basics.

Spring Tasks

Spring is unique in that the first task is always the same: <u>The First Spring Hive Inspection</u>. There should always be a goal associated with each task and this one is no different.

There is a primary goal and a secondary goal during the spring inspection:

- **The primary goal**: This is to see which hives survived the winter and to establish if they need to be fed or not. Winter can be long and cold. It's a possibility that your hives have nearly depleted their food stores and if so, they will need you to start feeding them if the weather doesn't allow for foraging anytime soon.

- During this time when you are determining their food status, you should also be checking their population size which is easily done visually as soon as you remove the hive cover. You should also check to see if there is any brood. A good hive will have lots of brood and a good population of bees due to having a laying queen. Brood indicates a future, growing population size.
- **Secondary goal**: This goal is mostly to see if hives are ready to split or not. Splitting bees is the process of growing more beehives from the ones that you already have. Beekeepers can split in the spring or the summer.

So how does knowing these things relate to being an intuitive beekeeper? We will continue to use the spring task as our example to answer this question.

First of all, we need to know when conditions are ideal for a first inspection. That part is easy: low wind, temperature ideally in the upper 50's to lower 60's and sunny to partly cloudy.

When this actually happens is something we won't know unless we keep an eye on the weather forecast. In my area I can get into my hives for the first spring inspection usually sometime in March or closer to April. At this point, maple has been blooming for several weeks and my hives have been growing.

What I know to expect from past experience is that my overwintered hives (*the survivors*) will have bees between at least 7–8 frames if not more, they will also have several frames of brood, meaning they have a good laying queen, tons of pollen packed around the brood nest and some capped honey leftover from last fall. Discovering these conditions is my primary goal of the first spring hive inspection.

If my hives do not look like this, it is a cause for concern and I must investigate further. As a side note, the majority of my winter losses are not from the cold but rather a high mite count. More on that later.

Intuition plays a role in all of this because I know that on the warm days, my bees have been taking advantage of the nice weather to gather maple pollen, thereby indicating to the queen to increase her egg-laying rate. She has been increasing the population of the colony for several weeks before I even do my first inspection of the year.

My intuition plays another part when I find a hive that is still alive but doesn't look like it's doing well which is usually indicated by a much smaller population of bees.

When you know what your colonies *should* be doing, *when* they're doing it and *what* their condition should look like, then you begin to understand what you're doing during inspections and *why* you're doing it. In essence, you develop a gauge for yourself to calibrate from and are constantly comparing your hives to the best and biggest hive you have at that time.

Gone are the days of looking in a hive without knowing why. This is how you build the confidence of knowing what to do and when to do it in beekeeping.

SUMMER TASKS

Now we are going to look at a summer task example and follow it through with understanding how intuition plays a role.

One of the most prominent things to do in summer is to harvest the honey crop. There are many factors involved with this particular task and it's right on point with knowing when you are having a honey/nectar flow in your area so let's start with that.

First of all, a honey/nectar flow is when a major nectar source is blooming, offering your bees the chance to make a surplus of honey. This surplus is what makes it possible for beekeepers to harvest honey from a beehive without taking too much of their stores. We will describe several elements of how you can tell when a honey flow has begun later in *Chapter 5: Working with the Flow*.

Depending on your area you can have anywhere between one to three honey flows per year. Generally speaking, a beekeeper can harvest from all but the last honey flow of the year, allowing to bees to keep enough food for winter survival from the final one. Also depending on your area, a single beehive can produce anywhere between 50–200 pounds of honey. To give you a visual this is roughly between one to four 5-gallon buckets of honey.

In my area, a honey bee colony averages 50+ pounds of honey. Sometimes a hive may surpass your expectations. For example, one of my hives produced 140 pounds of Wildflower honey to my surprise, outperforming all my other hives by far.

It's crucial to know when your honey flows begin and end. Having your beehives prepared before a honey flow is imperative in order to take full advantage of the flow and make as much honey as your bees can produce. Knowing when a honey flow ends is just as important so that you don't harvest too early because honey bees need to *'ripen'* the honey.

Honey is considered '*ripe*' when the bees have reduced the moisture content of the nectar down between 15–18%, generally speaking. Don't worry, you don't need to measure the moisture content of your honey although there is a tool for that. When a frame of honey is 75–100% capped with wax cappings, it is considered ripe enough for harvest. This is what gives you a nice thick honey that has a low moisture content rather than a high-moisture, thin liquid honey. With that basic understanding, let's go over the tasks involved in getting a hive ready for a honey flow.

First and foremost, a hive must always have a good laying queen, lots of brood and a large population in order to be productive.

When you have a beehive like this they will make you a honey crop. To get them ready for a flow right before it begins, you must assess that they meet these specific standards. With the hive cover removed from the hive body, you place a queen excluder on top of the brood chamber where the queen is located, followed by a honey super or two. After having replaced the hive cover, you're done and the hive is ready for a honey flow.

How do you know when a flow is over? When the primary nectar sources that provide the nectar have finished blooming. In my experience and research, a honey flow could last anywhere between 3–6 weeks depending on the plant species involved and the flow usually begins and ends with either the new moon or the full moon.

Intuition plays a role in summer tasks by understanding what a hive should look like during this season. A hive's population is at its peak during summertime.

Intuition also plays a role through observation of the plants that offer high nectar yields and their bloom time from start to finish.

In my experience, I see bloom times begin and end with the specific moon phases I just mentioned. If a bloom begins with the new moon, it ends or begins to taper off by the next new moon. It is the same if the flow starts with a full moon as well.

You will notice that intuition consistently plays a role in knowing what a healthy, productive beehive looks like at each point throughout the year and the transitions between each of them. The same applies to knowing the seasons in your area as you learn to see the relationship between your bees and the unique environment that they live in.

Now let's take a look at a fall task because it requires a little more insight and management than spring and summer do.

Fall Tasks

The element that makes fall tasks so unique is that you now have to consider several key points when working your hives:

- First, tidying up your summer tasks so that you can properly tend to your hives' condition is important to complete. If a hive is queen-less at this stage in the game it is imperative to correct this as soon as possible because there is very little time in the year remaining for a colony to raise a queen and produce enough brood to make it through the winter.
- On top of this, it's time to medicate for pests and disease again in a preventative manner to ensure your bees are healthy going into the winter, especially against Varroa mites.
- This is also the time to manage your hive equipment in such a way that your hives have the ideal amount of space they're going to need for winter. Too much space means that the

- bees have to expend more energy to heat up the interior of the hive.
- Your bees also need enough food to eat through winter in order to ensure that you have a strong, healthy hive to greet you at the first spring hive inspection next year.

These tasks bring us to another key for being a successful beekeeper:

- **Key #4**: **The successful beekeeper takes his or her winter losses** *in the fall.*

This key might need a little explanation. There will be hives in the fall that clearly will not make it through winter yet many beekeepers try to nurse them back to health regardless.

Part of your intuition will need to recognize those hives that simply will not repair in time. They are not only hopeless, depending on the circumstances, but they are also a waste of resources, ie the beekeeper's time and energy trying to force survival.

Here is an easy example: during this time of year in question, I go through each hive and assess their condition. If I find a droney hive, then I brush all the bees off of each individual frame and bring the empty equipment home to be cleaned up and prepared for spring.

In the meantime, those bees that I brushed from the droney hive will now become disoriented as they try to figure out where home is which is now gone because I took it away. As a result, they always end up joining the nearby healthy hives without issue.

So not only have I recognized a hopeless cause, but I've saved resources by giving the bees in the droney hive a chance to join another colony, adding to those populations and any honey that the droney hive had in it, I can now share with another hive that may not have enough food. This is also called, *'making the best of a bad situation.'*

The role that my intuition played in this case is based on past experience in trying time after time to fix a droney hive. Out of the many times that I have done this, only once was I successful however, it cost me three mated queens to do so. They killed the first two and finally accepted the third one. By the time they finally accepted the third, they were weeks behind in production compared to where they should have been. It was not worth the cost though it was worth the experience.

So now when I come across a droney hive, I remember that experience and do the best I can by doing exactly what I wrote above no matter the time of year it is.

Intuition is a constant juggling act, taking current elements into consideration based upon known-ideal circumstances. It can be fun or it can be frustrating but it is always worth it because you learn and apply, regroup and improve. Now let's take a look at winter tasks.

Winter Tasks

You may be asking what does a beekeeper do during the winter. And again, those tasks will vary depending on your climate. However, although the bees are more or less in a dormant state during the cold months, the beekeeper should have plenty to do. Note that there are no in-hive inspections during this time. Colonies should not be disturbed during the winter months. Disturbance causes physical excursion on the bees' part in reaction to a potential threat and that takes calories, meaning that the bees can potentially eat their winter stores more quickly due to the need to replenish calories to maintain hive warmth.

It's during the cold winter months that a beekeeper does two primary things:

- Fixes, repairs, assembles and paints any new or used equipment for the upcoming season, having it ready in time for spring (*remember Key #2: The successful beekeeper is always thinking one season ahead*).
- This is also the time to plan what you want to do with your bees in the upcoming year. It's time to ponder on your goals and work towards making them happen. This planning time will help make your seasonal tasks more clear as you work towards the bigger picture of your goals. Goals will change from time to time and that is perfectly okay.

Here's a great example of both of these primary tasks working together: After having harvested my honey for the year, I work on getting my hives in order (*a fall task*) and part of that is to work all of my hives down into a single deep hive body for the winter with one, full medium honey super on top; also known as a one and a half.

I live in a moderate winter climate, meaning that it doesn't get incredibly cold but we still definitely have a winter time and it is often wet.

Through experience, I have found that my hives winter more successfully in single deep hive bodies with a honey super on top versus double deep hive bodies or solely in a single deep hive body. Queen excluders need to be removed after the last honey harvest so that the queen can move up as her workers move up into their winter stores as needed so that she is not left in the deep hive body, blocked by the excluder to freeze during winter, separate from the warm, winter cluster.

As a result of working all of my hives into one and a halves, I have unused deep hive bodies and honey supers back at my place and can now work on repairs, repainting and cleaning up boxes and frames of excess propolis and beeswax so that everything is ready for the springtime. This task helps me to take inventory of the equipment that I have so that I can compare those numbers with my goals for next year. This tells me how much equipment I may need to build over the winter time in order to reach my goals.

Again, intuition is at play here because I'm thinking ahead. Intuition is all about what *might* be or what *may* be. Anticipation is a large part of your intuition, especially as a beekeeper. When you learn to anticipate, you sharpen your intuitive skills and it eventually becomes much more seamless as you practice and practice, always getting better throughout the seasons.

In covering these examples of seasonal tasks, it is my hope to illustrate what the various assessments may be for you throughout the year. Each season is unique and that is part of what makes beekeeping so challenging and so fun.

The next chapter is going to dive into more detail concerning how to utilize your intuition when diagnosing a hive based upon your in-hive inspections. It's the chapter that answers, *'well what do I do now?'*

CHAPTER 3

WHAT TO DO WITH YOUR ASSESSMENT
—Taking Action—

Have you ever found yourself in a situation where you had no one to ask advice from and ultimately had to make a decision all on your own whether it was right or wrong?

I certainly have found myself in this situation shorty after having finished my beekeeping apprenticeship and started keeping my own beehives. My beekeeping mentor was no longer available to ask questions from and there were some situations that were beyond my experience.

Ultimately, I had to use what knowledge I currently had acquired about honey bees and their habits and then do my best to apply it to the unique situation I was facing. Unfortunately, I cannot remember what that was at the moment and therefore cannot tell you.

Ideally we would all have a mentor to consult with when we need advice in beekeeping, however we are not all fortunate like that. It would be even better if our hives were always healthy and never needed preventative pest & disease management. Unfortunately consulting beekeeping books won't necessarily solve our issues either.

As a result, we are the ones that have to make the call when deciding how to handle a situation with a colony. Sometimes it's easy to know what to do:

- Too many hive beetles: put in beetle traps.
- Hungry hive: feed them.
- Accidentally squished a queen: re-queen them or let them raise a new one from their own brood or brood from another, stronger colony who can afford to share the resources.

Sometimes We Must 'Bee' the Doctor

Though there may be times when we are unsure of what to do, it is usually due to a couple of reasons; one reason may be that it is in fact a rare and unusual situation but the more common reason among beginners and not knowing what to do is simply the lack of exposure to enough beekeeping scenarios.

Hobby beekeepers usually have anywhere from 1–8 beehives and therefore do not need to work bees full-time or have the exposure to that kind of work like full-time beekeepers do with hundreds or even thousands of hives.

Fortunately, intuition can actually help us face unknown and unfamiliar circumstances in beekeeping when it comes to taking appropriate action steps during a hive inspection. Intuition helps when we have an understanding of how honey bees themselves work; in other words, their habits and ways of doing things. Books can actually help a lot here for knowing what to expect but we need to see that book knowledge in action in order for it to really stick in our heads.

An example of knowing about honey bee' habits is the propolis trap. We know that bees will fill in any empty space less than ⅜" narrow with propolis. So beekeepers came up with something called a propolis trap. Propolis traps look very similar to queen excluders but with smaller openings. In fact, the openings are too small for bees to pass through at all. So naturally when bees have an area like this they propolize it shut using propolis. This enables a beekeeper to come along, remove the trap from the hive and harvest the propolis in a very quick and convenient manner. This is an example of working *with* the bees, just as we will cover in *Chapter 6: Working with the Bees*. Great title right?

This is an example of how beekeepers can take advantage of the bees' ways of doing things and make it work for them. But it's also an example of knowing how bees work and that is the most important part I want you to understand here.

Here is another great example of something I do to use the bees' habits to my own advantage and to theirs as well. I utilize something I have dubbed *'Varroa Control Frames'* in my deep hive bodies that are used as my brood chambers.

Knowing that honeybees will fill in cavities larger than ⅜" by building comb, I use two medium honey super frames in each of my deep hive bodies. As a result, they will draw out comb on the bottom bar of those two frames to match the approximate size of the surrounding deep hive body frames to match. For some reason when the bees do this they almost always make it drone sized cells and the queen will come along and lay it with drone brood.

Varroa mites love using drone brood to propagate themselves in because the lifecycle of a drone has a longer incubation and developmental period, giving the mites longer to increase their own population.

So what I do is come along and check these frames at each hive inspection. If the bees have not drawn out the comb yet then I close up the hive, however if they have drawn out the comb on these two Varroa Control Frames and the queen has filled it with drone brood, I will scrape the comb off of the bottom of the bottom bars, thus removing all of the Varroa mites that are in it, and therefore physically removing hundreds of mites from my hives in one fell swoop.

When we understand how bees work, that very knowledge can help us to think of a solution to an unknown or unusual dilemma in a hive or even a common issue that we simply have not had experience with yet.

When it comes to having a colony with an issue that we must solve, some of the more common things we face as beekeepers are:

- A queen-less colony.
- A droney colony.
- Varroa Mites and their viral aftereffects.
- A hungry colony.
- Robbing.
- Equipment failure (*not an in-hive issue*).
- Weather conditions preventing us from accomplishing certain tasks on time and also assessing colony status.
- Overrun with Small-Hive Beetle.
- American Foulbrood (*horrible condition but fortunately rare in my experience*).
- Wax Moth damage.
- Bear attacks.
- Etcetera.

We are going to look at two of these more common issues that are more likely to occur and walk through them to demonstrate how intuition could play a role to finding a solution.

A Queen-Less Colony

Let's assume that we actually established that the hive is indeed without a queen. I say it this way because many beekeepers freak out when they can't find their queen and instantly assume the colony is queen-less, thinking they need to call every beekeeper they can to buy a mated queen for their hive.

I encourage all of my students to NOT search for the queen during each and every hive inspection. Rather, as beekeepers we need to look for *evidence* of a queen, namely eggs. The presence of eggs indicates a queen was in the colony within the last three days. There are hive inspections where we need to specifically spot the queen but not every single time.

When you look for evidence of the queen rather than the queen herself then you minimize the risk of accidentally killing her during inspections. You will learn more of this in the chapter explaining *The 7 Methods of Minimal Disturbance*.

When you come across a hive without a queen, it will look like one of these two conditions:

- The hive will still have a decent population of worker bees. This could potentially mean that the loss of the queen was recent. It could also mean that the hive recently swarmed, taking the primary (*older*) queen with them and now the bees that remain are raising a new one (*likely still a virgin at this point and difficult to locate*). This will be evident as a gap in the brood cycle. Eggs will most likely not be present because the old queen stopped laying for a few days in order to lose weight so that she could fly with the swarm. If this is the case then you should see recently chewed out queen cells still present, capped brood and possibly a virgin queen running around. There really isn't much to worry about if this is the case. If you see those cells, understand that there is most likely an unmated queen somewhere in the hive and that the hive will be fine. They just need time and very little disturbance from you for another two weeks, at which point you will want to come back and look for eggs again.

- The second scenario is usually the one that worries any beekeeper; the hive is queen-less and has a small, dwindling population. When I see these conditions it usually means that the hive has gone droney or they failed to raise a new queen. Sometimes the queen doesn't return to the hive after having taken a mating flight. A dragonfly could have snagged her out of the air for lunch or she was hit by a car while flying over a highway or something, you just never know. Whatever the reason that there is no queen, this scenario is the bad one. You're welcome.

In the second situation, there are potential ways of helping the colony get back on its feet. I would first analyze the resources they have: honey, pollen, remaining brood (*if any*), population size and even how many empty frames there are which tells how much resources they have.

In its weakened state, the colony needs your help to give them the time and resources it needs in order to survive. First, I would minimize the amount of space that the colony needs to protect, to keep warm and to ventilate. Therefore I would move them from the 10-frame box into a 5-frame nuc box.

This minimizes the area to a more manageable environment to suit their numbers. The entrance is smaller on nuc boxes and now they have a smaller entrance to protect which is ideal because less bees means that they have less *'man power.'*

I would also take any remaining brood and place it in the middle of the nuc box; brood is always structured in the middle in a healthy hive. Then I would take the frames with the most honey and place them on the outsides of the brood nest. Ideally there would be at least two frames with brood and at least two frames of honey.

If they did not have much food then I would take a frame or two from a stronger colony and give it to the weaker one. But the most important step is to get a nice frame of brood with eggs and young larvae from a healthier colony, brush the bees off of it into the donor hive and then give the frame to the weaker colony, placing it right in the middle of the nuc box. This gives them the brood needed to raise a new queen on their own and by placing it in the middle, the queen cells they draw out are in a better placement for protection and the added brood will boost their population.

You will know this method works and that they will raise a new queen from this brood frame when you return to the hive in 2–3 days to check for newly formed queen cells on the frame you gave to them. Intuition plays a role in this scenario in several ways:

- Recognizing the condition of the hive compared to what it should look like during this time of the year and also compared to your other, stronger colonies.

- Knowing that there should always be the three stages of brood in a healthy colony at all times: eggs, larvae and pupae (*pupae is always capped brood*) and the absence of one of the stages will alert you to having an issue.
- Recognizing the lethargic or possibly overprotective attitude of the bees upon removing the hive cover. The ability to read a colony like this comes with time and experience. Often colonies without a queen will be quite stressed and their attitude will show it by being more aggressive or '*flighty*,' meaning that they run about all over the frame quickly as opposed to a healthy colony which would simply continue to calmly work despite your presence.
- Knowing that a queen-less colony will always attempt to raise a queen when it has young larvae present; in this case it is given to them by the beekeeper from a stronger colony.

This chapter is about taking action once you have assessed that a colony needs your help. The action taken can sometimes be recognizing that you need to back off and let them do their thing as in the first scenario above.

Other times require more intervention and the juggling of resources like in scenario number two. Either way, you are taking action based on what your intuition tells you which is also based on the experience you're gaining and the knowledge you're gathering each and every time you work your bees.

You won't always be right and when you are it won't always work out but you must try still. Either way you are going to learn something useful for next time.

Robbing

Now let's take a look at a simpler but very frustrating issue that you will face each year. Fortunately it has an easy fix. Robbing is something that occurs heavily during a honey or nectar dearth. A dearth is simply when there is not a lot of forage blooming in order to provide enough nectar nearby for the colony.

This can happen in different ways; during the morning after having rain the night before, the nectar can be washed out of the blossoms, there will most likely be a lot of robbing that day. Plants will often produce more nectar by the heat of the day and the robbing ends temporarily. Or it can happen for entire days or even several days depending on weather and forage availability.

These are factors that we can actually learn to read and eventually predict within certain parameters after several years of observations in one area. The solution to robbing however is twofold:

- Employ entrance reducers to cut down on the amount of space the guard bees must maintain. Make sure to use the properly sized entrance for the warm season and the one for the cooler seasons as well, especially if you use solid bottom boards rather than screened ones. You do not want your hives overheating.
- If you see a hive being heavily robbed, just let it happen. It means that the hive being robbed was already weak and chances are that it would have eventually dwindled to nothing on its own. The good news is hopefully that the robbers are bees from your other hives so technically your honey will stay in-yard instead of feeding the neighborhood bees. As a side note, I have seen hives being robbed one day only to

be completely fine the next. I simply reduced the amount of equipment on the hive and gave them less space to guard.

In fact, when I see one of my hives being robbed by a mob of bees, I sometimes crack open the hive cover to allow it to be robbed even faster if I know the condition of that hive is hopeless.

I've seen videos and books recommending how to stop robbing by spraying water all over the bees to mimic rain so that they will return home or by spreading a sheet over the hive being robbed. If you have only one hive, go for it and try to save them. If you have multiple hives, learn to protect them before a dearth happens and accept it when it does. It's much easier on you. You cannot be there to protect them all the time, so just let nature be nature. Survival of the bee-ist.

Taking Action

Easily said but not always easy to apply, taking action is easier when you have taken the initiative to learn more about honey bee behavior through literature and especially experience. One way you could possibly gain more experience than what you currently have is by shadowing a beekeeper near you.

No matter what the issue is that we as beekeepers are trying to assess and resolve, the main take away and most important part of the assessment process is the final part; taking action.

This step is actually what trips up most people because they're worried about *'doing the wrong thing.'* Well guess what? You will do it wrong from time to time but that's how you'll eventually learn how to do it right. Plus when it comes to a large majority of beekeeping tasks, the bees are surprisingly capable of making up for many of the errors of their beekeepers.

Taking action simply means making a decision based on the best of your knowledge base in the moment. Taking action is the building block of experience and as you practice it, it will not only build your experience but your confidence as well.

Even at this point in my career twenty plus years later, there are still plenty of beekeeping mysteries that I simply will never know the answers to, however for the most part when someone asks me a beekeeping question I can answer it quite thoroughly and with confidence.

What this eventually morphs into is that you become your own reliable source of knowledge as long as you continue to educate yourself in the field of beekeeping and remain open towards learning more.

How to Take Action

I want to give you a brief step-by-step guide for taking action next time you come across a dilemma that you are unsure about how to proceed with. Let's break it down into three steps that you can apply as general steps to most situations:

Step #1: Identifying there is a problem—First of all, let's say that you have just assessed that something is wrong with your hive. This could be a whole lot of things but in essence it means that you recognize things are not ideal. Either way, you know that something is '*off*.'

Though you may recognize there is a problem, you may be unsure of what to do about it and in beekeeping when we find something wrong, we often feel that we need to solve it right away in order to avoid the worst case scenario. Part of this first step is to not panic or rush into action but rather to take your time and think things through.

Step #2: Resource assessment—This is where you need to rely on your current knowledge of how bees and the hive work and I don't mean manual work; what I mean is how they do things in such a way that you can use this to your advantage; i.e. their habits and social structure.

What resources do you currently have that could potentially help this hive's condition? Resources could simply be manipulating the hive itself by manipulating environmental circumstances (*like feeding*), or perhaps you have resources in your other hives that may be able to help the one in question. This resource could also be a more experienced beekeeper you know that can help you.

One simple example of using circumstances and bee knowledge to your advantage would be discovering that when there is a hive without a queen and does not have any hope of raising a new one.

The resource to help this hive would be a frame of young brood from another strong hive that can afford to share and giving that frame of brood to the problem hive in order to raise a new queen. Again, this is just a simple example among a greater scope of endless scenarios.

Step #3: Resource disbursement—After you have assessed the issue to the best of your ability and have determined what resources you have to potentially remedy the situation, it's time to implement it by taking action. Do it! And after you have done it, decide how much time you need to give the bees to do their thing before coming back and checking on their progress and to see if the issue has been resolved.

The best we can do is not necessarily fix anything but rather give the bees the resources they need in order to fix it themselves. Sometimes this works and other times it doesn't.

This is what being a beekeeper is all about. If you just have bees in a box and never involve yourself and hoping that the bees will simply thrive on their own, you are not a beekeeper, you simply have bees. You don't want to simply have bees, you want to *keep* them which leads to the fifth key to being a successful beekeeper:

- **Key #5: The successful beekeeper makes a decision and takes action.**

Even though there are times when I am stumped and simply do not know what to do in certain circumstances, when I take action based on my knowledge of honey bees and my action works out for the better, I am super excited and it's something that I will never forget should that issue ever arises again.

On the flip side, if the action I took did not work, I remember that too, meaning that now I know what NOT to do and what does not work. The awesome thing that happens when your action step did not work out is that the bees have a way of helping you to understand why it did not work and you have still learned something big.

Never be afraid to take action because it's what makes beekeeping so much fun. In the next chapter, you will read about how to anticipate your bees' needs. Knowing this is what helps you determine what actions to take and when to take them.

CHAPTER 4

Anticipating Your Bees' Needs

Anticipating your bees' needs is probably one of the more busy yet rewarding beekeeping tasks of the trade.

I love anticipating and preparing ahead of time even though it can be stressful when having lots of hives to tend to. Having fewer hives, ten or less, is very good practice for being able to manage more and more hives over time if you choose to. My mentor told me in the beginning that if I could keep ten hives alive and could do it well, then I could keep more after learning how to maintain those ten.

What does it mean to anticipate your bees' needs though? Don't they pretty much take care of themselves?

Well yes and no.

If bees took care of themselves then there would be no need for beekeepers, we would probably have men and women known as honey hunters instead, locating feral colonies and robbing their honey. In order to anticipate the needs of bees, we must become familiar with what their needs are and when they need them.

Do you remember the second key to successful beekeeping mentioned in Chapter 1? It's that the successful beekeeper is always thinking one season ahead. This directly ties into the elements that dictate your tasks as a beekeeper.

Anticipating your bees' needs will be an ongoing skill from year to year. I made a grave mistake one year in not anticipating my bees needs in conjunction with my management style at the time.

It was late summer transitioning into early fall when I had just finished harvesting my honey for the year and was preparing my bees for the winter time. I had begun feeding them sugar syrup to get some weight on them so they would overwinter well. As far as this example goes, I was correct in anticipating my bees' needs, making sure they had enough food for winter survival.

However, my management style at the time did not take into consideration the colonies efficiency. You see, at the time, I ran my hives in single-deep brood chambers. On top of the single deep I placed a queen excluder and then all of my honey supers on top of that in preparation for the honey flow. This is normal but I assumed that the bees were storing honey down in the deep hive body around the brood nest and therefore had plenty of honey for the colony.

I was wrong.

I later learned a very serious lesson about managing single-deep brood chambers: the bees, in all their glorious efficiency, realized that they had limited room for brood rearing and therefore used the entire brood chamber for brood only, not storing any honey in the deep hive body. After all, they had plenty of room to store honey in the honey supers on top…until the ignorant beekeeper (*me*) came along and took it all.

This left a very hungry hive with lots of mouths to feed and no provisions to do so. It also left a very perplexed and worried beekeeper who eventually learned how to properly manage single-deep brood chambers throughout the year in such a way as to properly prepare them for successful wintering. I now run my colonies in a standard one and a half. That is, one deep hive body and one medium honey super as the brood chamber having enough room for honey stores. They have more room to store honey for themselves and I harvest anything above that.

Observation Is Where to Start...

It can be a challenge to anticipate the needs of something that you know nothing about.

Iwer Thor Lorenzen, author of *The Spiritual Foundations of Beekeeping*, says that, "*By carefully observing such essential aspects of the bees' life one can come to see various dynamics at work that encourage further inquiry.*" p. 10.

Lorenzen is seriously onto something because he is basically saying that when a beekeeper observes something that he or she does not understand, then there is further inquiry to find an explanation to what they are observing. In this way, observations lead to questions eventually being answered. This is one of the many ways that we beekeepers learn.

Lorenzen goes on to say that in order "*...to understand the bee one has to step back and see the whole, not just as an integrated being, but as an integrated set in its wider context, the world through which it has its existence.*" p. 2.

This is especially important when understanding the bee in relation to its role as a participant in the plant kingdom.

Though you will see more in *Chapter 8: The 7 Methods of Minimal Disturbance* about anticipating your bees' needs, it definitely warrants its own chapter here.

Looking at the larger picture, anticipating your bees' needs begins with the seasonal needs, always preparing one season ahead. Within that seasonal framework are individual seasonal tasks that must also be anticipated and prepared for.

There are certain tasks that all hives must undergo and then there are other tasks that only need doing depending on the beekeeper's goals for his or or specific operation or a hive's individual needs and circumstances at that time.

Let's talk a little about the specific tasks that every beekeeper must do with all of their beehives.

What Must Be Done for All Colonies

No matter who you are, where you and your hives are located geographically and no matter how many beehives you run, there are definitely certain seasonal tasks that must be done. Let's cover some of the more common ones now:

- **Pest & Disease Management.** By far this was the most intimidating one for me when I was apprenticing. Remember that I mentioned I did not know what to do, when to do it or even how? However, the bees did not care about my lack of knowledge and eventually succumbed to hive beetles and heavy Varroa mite loads, ultimately dying. Pest & Disease Management is something that needs to be done on-time in a preventative manner, otherwise it's generally too late to treat for an issue once a hive is already showing major signs of infection or infestation.
- **Supering for a honey flow.** It is crucial to give honey bee colonies enough room to grow as their population increases

and as their population does increase, they will have more foragers available for bringing in nectar, thus having the need for more room to store the surplus honey during a honey flow. It's already been proven mathematically that a queen has enough room to lay in a single deep hive body, so I will not go into that in this book. However, you will need to give your bees enough honey supers to store honey in during a flow; for them and for you.

- **Honey must be harvested.** One of the most common things I hear from my students is that they want beehives for the environment and for pollination, not necessarily the honey. They have a fear of taking too much of the bees' food. However, harvesting honey is a seriously major part of managing beehives or else they may become honey bound and swarm. When a colony feels cramped and crowded, like when their hive is packed with honey from wall to wall (known as being '*honey bound*'), they begin to develop the swarming tendency and will end up taking much of the honey and most of the bees as they fly away, out from your equipment and begin to search for a new home somewhere out in that big wide world far away from you.
- **Hives must be prepared for the winter.** Granted that some areas of the world have extremely light or mild winters, honey bee colonies must be ready in some degree for the winter nevertheless. Your location and the intensity level of your wintertime will dictate approximately how much honey your hive will need to survive through winter and what measures must be taken to winterize your colonies such as wrapping them or

not. It is your job to analyze colony conditions in the fall as you help to prepare them for the winter for your specific climate.
- **All beekeeping equipment must be maintained**. The majority of beekeeping equipment is woodenware and it must be maintained. Wood rots due to exposure to the elements. Hive equipment must be painted or at least treated with something fairly waterproof to protect if from the elements. Each winter, I gather all of my equipment that is not being used on bees and clean it up for next spring. This task often includes repainting deep hive bodies and honey supers. Aside from helping my hive bodies and honey supers to last longer, the fresh paint helps me to maintain a nicer looking operation which is important to me because my apiary locations are all on other peoples' property. Nice looking, well maintained apiaries will go a long way in maintaining good relations with gracious, location hosts.

The moral of the story is that you cannot simply *have* bees, you need to be involved and get your hands sticky so-to-speak. This is bee-*keeping*.

In anticipating your bees' needs and preparing yourself ahead of time, this is indicative that you must also understand the equipment that is used in beekeeping and how to manage it ideally.

Managing Hive Equipment

Managing your hive equipment is a part of anticipating your bees' needs. They need a brood chamber in which to develop and raise their young. That equipment must be full of frames, whether it's

8 or 10 frame equipment, with your choice of foundation: BPA-free plastic, wired wax or none at all, allowing the bees to build their own within a bare frame.

Honey bees will need to have a place to create their brood chamber and also to store abundant amounts of honey at the appropriate time. You must have a certain amount of honey supers per hive and each of those must also be full of frames as well.

Each colony must also have a hive cover and bottom board that are in good condition that do not leak and are not full of holes or loose materials. Hive covers keep the rain out and therefore help to maintain the colony's warmth.

Bottom boards bear all the weight that a hive may consist of, up to a couple hundred pounds at times. As you can imagine, though underestimated and not thought of often, a bottom board has a huge job and must perform at all times while in use, especially since they also function as the hive entrance.

Another part of managing your hive equipment is when it is *not* being used, it must be stored and stacked in an area out of the elements. Keep your hive covers stacked together, honey supers stacked on top of one another and your queen excluders stacked somewhere they will not become damaged.

I have met a lot of beekeepers and have never seen a sloppy beekeeper as a successful beekeeper. Their sloppiness in equipment management was simply an indicator of how they managed their beehives…not very well.

So be a tidy beekeeper and possibly even an organized one and you will be developing and maintaining a habit that will go much further in life than just helping to have better beehives.

Frames that are not kept in boxes and perhaps are laying around the bee yard or honey house and kept in disrepair is such a waste and can be prevented. Wooden frames with wax foundation will need the foundation replaced if not kept stored in their respective boxes. Plastic frames warp in the sunlight and heat, becoming unusable.

The habit of tidiness and organization goes beyond equipment and also applies to your valuable beekeeping tools as well. Keep your hive tools, smoker and protective clothing in the same area when you're finished using them. This way, they will always be where you expect them to be when the time comes to use them again.

In summary, the key things to take away and cultivate in your beekeeping practices when it comes to anticipating your bees' needs are these:

- **Colony needs**: Know what resources they need to maintain a thriving colony and approximately how much those resources fluctuate throughout the season as the colony population increases and decreases.
- **Environmental circumstances**: Begin learning the seasons in your area; how long they are, what forage is available when it comes to major and minor sources of nectar, how long your honey flows are and what your specific climactic challenges are.
- **Colony management**: Learn what tasks must be done for every colony throughout the seasons and do them on-time (*which means preparing ahead of time*).
- **Equipment management**: Learn how to manage your equipment while it is being used on your colonies and learn how to maintain your equipments' integrity when it is not in use.

Following these guidelines consistently will set you up to find the success you want as a beekeeper and who knows, you may find yourself mentoring a beginner in your area to whom you could be an example as well.

CHAPTER 5

Working with the Flow

Paying attention to nature; like the seasons and plant bloom as honey bee forage comes and goes in availability, is what takes an average beekeeper to being an wise beekeeper.

In *Chapter 4: Anticipating Your Bees' Needs,* your mind was being prepared to think about how you can anticipate the appropriate tasks ahead of time in order to prepare for them. This chapter is about one of those more exciting and important tasks in a beekeeper's year; the honey flow.

What is especially unique about this aspect of beekeeping is that the elements of a honey flow vary from region to region which is how we get different varieties of honey. Your area and climactic conditions will have some plants similar to the larger, surrounding area and some that are unique only to your specific area.

Easily one of my all-time favorite books that I recommend to every beekeeper is *American Honey Plants* by Frank C. Pellett. This book talks about major and minor sources of forage specifically for honey bees and it even breaks things down into individual states so that you can see what's available in your area more or less.

Major sources of nectar are what will create and provide a honey flow for your bees and it also means that you will have something to harvest for yourself and others. Minor sources of nectar do something that I call, *"bring home the groceries."* It's enough to maintain your colonies needs but not enough to produce a surplus of honey for harvest. It's kind of like living from paycheck to paycheck.

One of my first goals when moving into the north Georgia mountains at the foot of the Appalachians was to learn and identify which plants bloom, when they bloom and for how long. I learned the major sources of nectar and discovered that in my area we have three honey/nectar flows, two of which can be harvested from (*spring & summer*) and the third (*which is unreliable from year to year*) is one that the bees must keep for themselves in order to have enough winter stores.

Our spring flow pretty much begins in May and lasts for about a month. A couple weeks after the spring flow is over is when our summer flow begins which runs through the end of July. So I harvest a spring wildflower honey first and then sourwood honey at the end of summer and that's it for the year.

These sources are unique to the tri-state area that I live in and you will have plant varieties unique to your area as well.

Honey Flows—What to Look for and How to Know When It Starts
- **Plant Life**

First of all, you must become aware of the plants that offer the *major* sources of nectar in your specific area. Frank C. Pellet's book is one way that you can do this. Talking with local beekeepers is another way to know with one catch; not all beekeepers understand the plant world in relation to beekeeping. So there's that.

Once you have become aware of what the major sources of nectar are, learn to identify them. This means that you need to go on walkabout.

For me this meant that I needed to get some plant identification books. You can find a ton of them online. Specifically look for books that cover trees and another book for wildflowers in your area. Then get to work!

Identifying plants is going to be easier during times of the year when the leaves are still on the trees, shrubs and of course the flowers are the easiest way to identify something. You will eventually get so good that you'll recognize them even in winter when the leaves are no longer on the plants believe it or not.

Keep in mind that the plant you're looking for may not be in bloom yet or it has already bloomed this year and is finished. Do not let that stop you from identifying it though. When you can observe honey bees on certain floral sources, this is the best time to identify that plant and also what color the pollen is from that particular pollen source. For example, blackberry pollen is somewhat grey in appearance. Knowing when blackberries bloom and seeing my bees bringing home grey-ish pollen tells me that blackberries have begun to bloom even before I realize it.

- **Whitening of Comb**

When sufficient resources are coming in, the colony begins to build out new comb and fix old comb. As they do this, the comb has the appearance of being '*whitened.*' What this means is there is fresh, newly created beeswax that the bees are producing from their abdomen using the incoming nectar as a resource to do so.

- **Building Out Bare Frames**

Bees generally will not draw out comb on new, bare frames whether it's wax foundation, plastic foundation or foundation-less frames. However, when they have the abundance of resources coming in during a flow, their population increases and those two factors will stimulate growth in the hive. To make room to store the incoming nectar they will draw out comb on bare frames at an alarming rate. If you poke your face in a hive during the honey flow you should expect to see the colony building out comb and filling it with nectar as fast as they possibly can.

- **Increased Weight of a Hive**

Incoming nectar means that the hive is about to put on some weight...actually a lot of weight! A beehive can produce 50–200+ pounds of honey in one honey flow, so be prepared to bottle some honey. You can gauge the weight of a hive by standing behind it and lifting it up slightly from the bottom board. Do this once a week during a flow and feel the difference and remember to give them enough room (*plenty of honey supers*) during this time.

- **Aroma of Ripening Honey**

There is nothing like the aroma of ripening honey in an apiary. Usually you can smell ripening honey in the evenings while standing in or very near your hives as the bees are coming home for the night. Some honey varieties are more noticeable than others. Goldenrod for example kind of smells like dirty socks. I don't know why this is the case and I also cannot explain why I like this smell...so I won't.

- **Bloom**

And of course the most obvious sign that you are having a honey/nectar flow is by identifying the major sources of nectar in your area when they are in full bloom. Usually a honey flow begins gradually up to a point, then it hits hard. Then after about 4–6 weeks it will begin to taper off and come to a end for the year.

Not every tree of the same type blooms at the same time. Some will begin to bloom one day, then others will follow. Another interesting observation by my mentor was that the bees worked the hardest and brought in the most nectar on hot, cloudy days rather than full sunny days. Watch what happens in your area from year to year and be one with nature.

Something that I have noticed in my area is that a flow will begin gradually leading up to the full moon which is when it hits hard. The flow vigorously continues until the next full moon, at which point it tapers off over the next week.

Working with the Flow

Identifying the honey/nectar flow is one thing but you also need to know how to work *with* the flow for your bees' needs and so that you can make the most of it as well.

Working with the flow is much simpler than you may realize and is actually one of the easiest things to do in beekeeping once you know what to do. Of course, after having gone through one honey flow, you will be much more prepared for the next one in understanding what to do.

Up to this point in the season you have been making sure that you have strong, healthy hives with large populations and a high

quality, laying queen. Only strong hives will make excess honey during a flow. Your job is twofold:
- Make sure that you have 2–4 honey supers for every hive you expect to make a honey crop. Weak colonies may not make surplus honey for you to harvest but the incoming resources may be enough to get them back on their feet during a flow.
- Super your hives for the honey flow. Supering a hive literally means placing honey supers on your hives for the purpose of making honey.

There are several methods or strategies for supering a beehive but in its simplest form, just put the supers above the brood chamber(*s*) and if you're using queen excluders (*highly recommended*) then place them above the excluder. If you are starting off with completely bare frames (*without comb or drawn beeswax*), only place one box on for the beginning of the flow and add another once the first is 75–85% full.

If you are starting off with a super with drawn out comb, place two supers on at the beginning of the flow and judge if you need more after a couple of weeks into the flow.

It's more or less that simple. You can keep adding supers on top as long as the bees are still bringing in nectar. Pay attention to the bloom so that you know when it has ended. Always use your boxes with drawn out comb first if a larger crop is your goal. Undrawn frames require more work for the bees to fill, though they make it look like an easy task.

You can also under-super. Under-supering is when you have one honey super that the bees have almost completely filled with honey.

Remove that super and place a fresh one in its place, then replace the original honey super on top of the new one. The bees will begin working on the newer super much more quickly, filling it with honey and eventually finishing the original honey super soon afterwards.

As you experience each flow and the signs to look for to know when it hits, you will easily anticipate it from year to year naturally.

CHAPTER 6

Working with the Bees

Working with the bees. Wow, it sounds pretty obvious that working with the bees is what we do as beekeepers right? Well, yes but that's not what I mean when I say that we need to work with the bees.

Working with the bees means that you understand their habits, instincts, behaviors, how they do things, why they do things and when they do things which all ties into their social structure and how they interface with their external environment. Then you take this knowledge and manage your goals by utilizing beekeeping equipment to your advantage season by season. Let's go over some basic practices before getting into specifics. Here are some key points to consider when working with honey bees:

- **Respect their space**: Honey bees are highly sensitive to disturbances. Avoid sudden movements when working around them to prevent triggering defensive behavior. Wear appropriate protective gear, including a veil, gloves, and a bee suit, to minimize the chances of being stung, according to your comfort level.
- **Study their communication**: Honey bees communicate through complex dances and pheromones. Learn to interpret

their behaviors to understand their needs. For example, the waggle dance conveys information about the location of food sources. Seeing this is generally a sign of a thriving colony and means there are flowers blooming for them nearby.

- **Provide a suitable environment**: Bees need a healthy and diverse environment to thrive. Ensure they have access to abundant nectar and pollen sources throughout the year. Wild cultivated environments are preferable rather than large groves and orchards where there are not only fewer pollen varieties but higher pesticide use as well.
- **Regular hive inspections**: Conduct regular inspections to monitor the health and productivity of the colony according to seasonal tasks at hand. Look for signs of pests & disease and implement preventative measures accordingly. Carefully handle frames and avoid crushing bees. Document your observations to track the colony's progress over time if that suits your personality.
- **Practice responsible beekeeping techniques**: Implement sustainable beekeeping practices, such as using organic treatments for pests and diseases, avoiding chemical pesticides in the vicinity of hives, and promoting natural hive development. Also, you may consider becoming a certified beekeeper to stay updated on best practices so that you are always learning

Remember, working with honey bees requires patience, attentiveness, and continuous learning. By respecting their natural behaviors and providing a supportive environment, you can foster a harmonious relationship with these remarkable creatures while promoting their well-being.

Here's a non-beekeeping example of working *with* things for the sake of simplicity: Having looked at the weather forecast you know that it's going to rain tomorrow. Knowing this you decide to move the plants on your porch out in the yard so that they will be watered when it rains so that you don't have to spend the time to water them by hand. Let's make a mathematical equation out of this:

Knowledge/Experience x Need = Advantage

Now let's look into some beekeeping examples and apply this formula to them because this is after all a beekeeping book. Let's say for this example that you have a lot of fresh, brand new frames with bare plastic inserts or wax foundation that you want the bees to drawn out with wax.

The bees will not perform this task just because you put a fresh box on them filled with new frames. This action requires resources and honey bees are incredibly smart when it comes to expending and utilizing resources. The behavior of a honey bee colony is to draw out beeswax under two natural conditions most of the time:

- After having swarmed, they gorged their bellies with honey in order to create wax from their abdominal glands in order to begin building comb at their new home location, wherever that may be. The reason for this is so that the queen can begin laying as soon as possible for the survival of the colony, and also so that they have a place to store nectar and pollen.

- The second condition, and the main point of this example, is that honey bees primarily draw wax only during a honey flow. So naturally, to fulfill your want for the bees to draw out your new frames with beeswax, you will need to manage your equipment to do this *during a flow as per the bees' natural behavior*. It is during this time that plentiful resources are coming in and the bees feel that they can afford the resources for growth of this nature.

Let's apply the equation above to the last example:

Knowledge/Experience—Knowing that the bees will easily draw comb during a honey flow.

X

Need—The need to have your new frames drawn out with comb.

=

Advantage—The bees draw out the frames quickly *and* fill them with honey in the process, just as you wanted.

One of the more common ways of relying on the bees' behaviors to your advantage is when you find a hive that seems to be doing fairly well population-wise, and possibly resource-wise as well but for some reason it looks like they lost their queen recently for whatever reason.

Rather than freak out and order a queen from a beekeeper near or far you could simply give them the proper resources to make their own queen within a couple of weeks.

This is done by taking a frame of eggs and larvae from a stronger colony, shaking or brushing the bees off, and then transferring the frame of brood to the colony in need. They will very quickly begin to draw several queen cells and raise a new queen on their own. And this knowledge also plays a huge role when it comes to making splits in beekeeping.

BEEKEEPING: A SYMBIOTIC RELATIONSHIP

The relationship between honey bees and their beekeeper is a remarkable symbiotic bond built on mutual care and reliance. Beekeepers strive to provide a safe haven for the bees from predators, weather, pests and diseases, creating unique hive structures and maintaining optimal conditions for their colonies to live within. In return, the bees, astonishingly, live in manmade boxes and gift the beekeeper with the nectar of their labor, honey.

Beekeepers attentively tend to their bees, monitoring their health, managing pests, and ensuring not only their survival but their ability to thrive as much as honey bees can. They extract honey with precision and hard work, ensuring minimal disruption to the hive. This partnership is a delicate dance, where the beekeeper respects and nurtures the bees' natural instincts, fostering harmony and sustainability for both the bees and the keeper of their sweet golden treasure.

There are seven points I want to define for you regarding the symbiotic relationship between beekeepers and their bees:

- **Partnership:** The relationship between beekeepers and their bees is founded on a partnership of mutual benefit. Beekeepers provide a safe and nurturing environment for the bees, while the bees, in turn, offer the beekeepers the gift of honey and other valuable products of the hive. This in turn benefits the community as a whole as well, not just the bees and their keepers.
- **Care and Attention:** Beekeepers are deeply invested in the well-being of their bees. They closely monitor the health of the colonies, regularly inspecting the hives, and taking preventive measures to ensure the bees' well-being. This includes managing diseases, pests, and environmental factors that may impact the bees and taking action in each of those cases.
- **Hive Management:** Beekeepers are responsible for maintaining the hives and ensuring they are in optimal condition. They construct and repair hive equipment, provide sufficient space for the bees to grow, and facilitate efficient honey production of high quality for the community, as well as manage hive resources among the colonies to help the ones in need.
- **Harvesting Honey:** Beekeepers carefully extract honey from the hives, employing methods that minimize disturbance to the bees. They utilize techniques such as smoking the hive to pacify the bees, allowing them to remove excess honey without causing harm and of course this should all be done in a sustainable manner.
- **Sustainable Practices:** Many beekeepers adopt sustainable practices that prioritize the long-term health and survival of the bees. They avoid using harmful chemicals or pesticides

that could harm the bees or contaminate their honey, opting for natural and organic beekeeping methods instead.

- **Education and Research**: Beekeepers continuously strive to expand their knowledge and understanding of beekeeping. They actively participate in research, attend workshops, and engage in information sharing to enhance their skills and contribute to the natural understanding of bees and their behavioral roles in the environment.
- **Respect for Nature**: Beekeepers recognize the integral role bees play in our ecosystem and the importance of preserving their natural habitats. They promote biodiversity, support pollinator-friendly initiatives, and advocate for environmental conservation to safeguard the well-being of bees and their vital role in pollination.

CHAPTER 7

Queen-Keeping

No one likes failing or sucking at something and it would be great if we could always do something right the first time that we try it. The unique thing about us humans though is that with practice we have the potential to get really good at doing stuff.

Beekeeping is no different. The more exposure you have to bees, the more you will learn as you become familiar with the honey bee colony, its in-hive intricacies and the various tasks of the beekeeper, especially if you are always approaching a set of circumstances unique to each hive with the question, '*why*'? When we ask questions, answers will come.

It is always easier to learn from others' mistakes and apply them to what you're doing but don't be afraid to make mistakes, because you will. As long as you're learning from your mistakes and adjusting your approach for better results the next time, then you are well on your way to becoming a beekeeper rather than a bee-haver. Anyone can have bees but it takes a diligent person to keep them.

So why is this chapter called 'queen-keeping?' Beekeeping is a broad term if you really think about it. It is the conglomeration of all the tasks of a beekeeper to care for the entire colony as a whole.

The Queen's Role & Importance

The queen has the most crucial role in a honey bee colony, responsible for laying eggs and maintaining the overall attitude, health and productivity of the hive. A quality queen possesses specific qualities that are essential for the survival and success of her colony.

One of those qualities is her ability to lay a large number of eggs. I've read that she can lay anywhere from 2000–3500 eggs in a day but I have never actually counted because why would I do that? Laying so many eggs in a day ensures that the colony is constantly refreshing and replenishing enough worker bees to gather the resources of nectar and pollen, care for the developing brood and protect the hive from would-be predators.

Another quality of the queen is the ability to control whether she lays a fertilized egg to become a worker bee or an unfertilized egg to become a drone. She can choose which of the two that she lays and generally this is going to be according to the size of the cell that she is inspecting. A worker-sized cell will receive a fertilized egg while the larger, drone-sized cell will receive an unfertilized egg.

Among her many talents, her ability to produce pheromones help to regulate the behavior and development of the members of the colony. Her pheromones can influence the bees' foraging, behavior, attitude, temperament and grooming.

A good honey bee queen is also able to maintain a healthy and disease-free colony by laying eggs in a clean and hygienic environment, and by producing pheromones that signal to the workers to remove dead or diseased brood. In addition, a good queen is able to detect and respond to potential threats to the colony, such as invading predators. If you have ever tried to search for a queen in your hives then you know that she is very good at hiding.

Finally, a quality queen is able to adapt to changing environmental conditions very well, mostly if they are natural rather than human-influenced by transporting hives from one climate to another. Within her ability to adapt the in-hive conditions based on external environmental conditions, she can regulate the size of the colony by adjusting her egg-laying rate and even lead the colony into the swarming behavior.

In all of my personal experience, when the queen is healthy and doing her thing, then the entire colony is thriving. If we would focus on the well-being of the queen then the health and well-being of the colony will naturally follow as a *result* of that concentrated effort.

The health and vitality of our queens result in the level of quality they have and perform with. There are several ethical issues in determining or defining what a 'quality' queen is depending on who you ask. If you ask science, a quality queen is one that is selected and manipulated into being for her particular genetic traits. These traits usually include docility, overwintering ability, foraging habits and hygienic behaviors. A queen like this is often made in a lab so to speak, through the practice of artificial insemination.

If you seek the opinion of what a quality queen means to a naturalist beekeeper, you hear a very different perspective. Rather than artificial laboratory methods of creating a 'super-bee,' a beekeeper who prefers natural methods of beekeeping is going to prefer natural selection through the natural breeding process of allowing a queen to go on a mating flight, also called a nuptial flight, mating with 10–20 drones and allowing the elements of genetic diversity and locally adaptive bees to be what they consider the best quality of queens; i.e. survival of the fittest.

In case you have not figured it out by now, this author is in favor of more natural methods of keeping bees and maintaining a healthy, broad diversity in a locally adapted stock of honey bees.

There are several elements and colony characteristics that we can look at in order to determine if we have a quality queen or not. Some of them we have control over and the rest of it we do not; and that is perfect. Let's go over these elements in detail now:

- **Minimal disturbance.** In helping to maintain a quality queen, it is important to disturb the colony in the least way possible. This correlates directly with one of the Methods of Minimal Disturbance you are going to learn about in the next chapter. One of the best practices that we as beekeepers can maintain is looking for evidence of a good laying queen rather than trying to physically locate her every single time that we perform a hive inspection. It is not always easy to locate and spot a queen even if she is marked, and the longer we are in a hive, the greater risk we put her in by rummaging around and manipulating frames in and out while trying to find her. Instead of trying to spot her every time, learn to judge her quality by 'reading' the hive conditions using the elements listed below.
- **Look for eggs.** Eggs are almost a sure sign that there is a laying queen present. I say *almost* because there is the possibility of a laying worker. When you see a single egg in a cell (*rather than multiple eggs*) and there is a fairly large area of the frame laid with eggs, bordered by young larvae, bordered by older

larvae and possibly even capped brood, then there is a really good chance you have a quality queen present in the colony. This point borders with another sign of having a quality queen present; also having a nice brood pattern. Queen honey bees tend to lay in an almost rainbow pattern across the frame. Look for this and also look to see if the brood is *'rainbowed'* by pollen and open nectar in the upper corners of the frame as well. These are signs of a healthy colony and a healthy colony means that they have a nice queen. When you see eggs, it means that a queen has been present and laying within the last three days because a honey bee egg hatches on day four.

- **Check the overall demeanor of the colony.** A stressed hive has a certain buzz, meaning that they could be queen-less, low on resources or trying to raise a new queen. On the opposite end of the stick, a healthy colony could almost care less that you're there. You will notice that even as you remove individual frames for inspection that they carry on with their work, without any sign of being disturbed. The attitude of a colony will tell you a lot about the status of their queen.

- **Productivity.** A productive hive is a queen-right hive. If you ever see where the bees are whitening the comb to freshen it up as they build out and restructure damaged comb, this is the sign of a colony with a healthy queen. One of the biggest tells of an incredibly awesome colony with a quality queen is usually seen in the summer months on the face of the hive and that's washboarding. Washboarding is when a lot of honey bees are gathered on the front of the hive, moving in unison back and forth, back and forth. It almost looks like they're scrubbing the deck with their mandibles. I do not entirely understand what they're doing, but I have only seen strong colonies perform this action.

By paying attention to our queens by way of looking for the qualities and evidence that make a good queen, we are able to be more mindful in our practices to maintain beehives that thrive.

In essence, the queen possesses several very unique and vital qualities that make her role in the colony of the utmost importance. Her lifespan compared to that of drones and workers demonstrates her essential role in producing a highly populated colony for the purpose of survival. Her pheromones make this especially unique, basing the function and success of the colony on smell. And the fact that she can mate with multiple drones and have enough sperm for the remainder of her lifespan is astonishing.

The success and survival of the queen for the sake of the colony cannot be overstated, nor can the importance as beekeepers to place her at the center of our focus when it comes to maintaining our precious bees.

So if queens are so important, should we concern ourselves with raising specific genetic stock in our apiaries? And if so, what is the best choice? Is it Italian bees, Saskatraz, Carniolan?

Each unique bee has certain qualities that beekeepers have found to be desirable. These traits range between but are not limited to:
- How quickly a colony builds and grows in the spring time.
- The colony's gentleness.
- The amount of honey a colony can produce or even how often they get the swarming tendency.

Which one is right for you? Well here is a thought that will hopefully help clarify things for you. Let's say that you have decided to purchase purebred, Russian queens to re-queen your apiary's colonies. The queens have finally arrived in the mail and you anxiously go pick them up and then get to work introducing them into each of your hives in order to raise bees with all of the qualities that Russian stock has to offer.

A few days later you go back to make sure that each of your new Russian queens have safely made their way out of their queen cages and have successfully been introduced into their respective colonies. Then a week later you come back, hoping to see that the queens have begun laying eggs and indeed they have! How exciting! You are now well on your way to having Russian bees as the new eggs quickly develop into larvae, then into pupae, eventually emerging as fully grown adult Russian worker bees.

A few weeks go by and it's time to perform hive inspections for whatever reason. During the inspection you discover swarm cells or supersedure cells, you're not really sure but what you do know is that they are replacing one of your really expensive Russian queens. You tell yourself it's okay because the new queen will simply mate with your new Russian drones and therefore continue the Russian stock you so meticulously try to maintain.

The day comes when the new queens emerge and fight amongst each other till only one remains…the strongest of the queens! Within three days, this new virgin queen, a daughter from your original Russian queen leaves the hive on a mating flight. She lifts off from the landing board and flies off to where only bees know to go for such naughty games about half a mile away.

What you don't know and won't realize for some time is that half a mile away are where your long-distance beekeeper neighbor's Italian drones gather, looking for young virgins to play with. Your Russian queen returns successfully to her colony, pumped full of Italian drone semen, and within a week or so begins laying Russian/Italian eggs.

There goes your endeavors to maintain and run Russian stock. Now what?

Well unless you live on an island then it would be in your best interest to learn from this lesson rather than go and buy more Russian queens.

This illustration demonstrates that while your intentions are sound, mother nature has other ideas that you have absolutely zero control over.

This brings us back to one of our original questions: Which is the best honey bee genetics for you? Here is my answer: the best honey bee genetics that you should maintain in your operation is a genetically diverse, locally adaptive strain to your specific area.

What does this mean? It means that the colonies that survive your winters are hardy and tough. It means that *if* you buy queens and re-queen any of your hives that doing so with locally raised queens is a must. It means learning enough about bees so that you can create the circumstances under which your hives have the ability to raise their own queens. It means increasing your hive numbers by catching local swarms and housing them in your own hive equipment is a good thing.

I am asked fairly often by my students what specific genetic stock I run and I always tell them that is the least of my concerns. Then I tell them what I just told you, that maintaining hives from my winter survivors and catching swarms is the stock that I run; locally adapted and genetically diverse.

I realize that it's a cool idea to have a specific stock but it's even cooler to not have one more thing that's out of my control to worry about. There's plenty of things in beekeeping to give my attention to that I do have control over. It's a relief to know that I don't have to worry about the genetics of my honey bees on top of everything else.

Russian queens are amazing…in Russia.

Try to be a queen-keeper and learn to identify the signs of a high-quality queen in a hive rather than trying to find her physically.

As you gain more experience and see more hives throughout the months and years, you will be able to judge a hive based on the sound the hive cover makes when you pry it up and off. In general, the hive cover on a kick-ass hive will pry off with effort and difficulty, making a popping sound as it releases from the top bars (*if you don't use telescoping covers, in which case the inner cover is your indicator*). A hive cover that pops off with ease, making virtually no sound is most likely a weaker hive.

The reason for this is because a healthy hive propolizes everything together, making it stuck in place. A weak hive does not have the resources to do so.

Taking Your Time

It's all about intention. When you choose to do something intentionally rather than just going through the motions or rushing to finish something, the results will be of a much higher quality and you as a beekeeper will be a much better beekeeper, choosing quality over quantity. Intention trains you to be more effective in your practices and your bees will benefit from it.

Enough cannot be said about taking your time in beekeeping. More specifically, taking your time while doing in-hive work is the important part. This is most obvious because of the fragility of the tiny little bees but especially because there's only one queen in the hive and she must be protected.

I have had my time of rushing through hives just to get the work finished. I did not enjoy it and beekeeping is something that should be enjoyed and experienced fully.

The urge to rush through a hive generally comes from one of two reasons for beginners:

- You are nervous having a hive open, being exposed to thousands of potential stingers **OR**
- You feel overwhelmed, not knowing what to do or even what you're looking at.

Taking your time can help with both of these issues. By taking your time during a hive inspection, you expose yourself to the bees but in a way that allows you to observe them at work, allowing you to learn from what you are watching.

Remember, each beekeeper should suit up using protective gear according to their individual comfort zone for being around bees. If you are so nervous around bees that you rush through working them or avoid working them at all then you need to have a full-coverage suit. Then when you are doing a hive inspection it is important to remind yourself that you are protected. The bees can't sting you if you're properly geared up. So get used to this feeling of safety and protection and use it to your advantage to observe, learn and ultimately enjoy what you're doing in the moment. This is your time.

Also get really good at lighting a smoker and keeping it lit. Even the most docile of colonies has a bad day sometimes just like you and I. A smoker acts as your second line of defense and will make the difference between a good experience and a traumatic one.

If you are feeling overwhelmed because you do not completely understand what you're seeing or what you're doing, that's okay. Time and exposure will build towards your experience. There are many resources available through books and online that can help you understand how to identify things in the hive.

I remember the feeling of having bees in the very beginning of my beekeeping career, excited yet oblivious of what to do. I did not like that feeling. This is where taking the initiative to educate yourself comes into play. In addition to literature or online resources, try to find a beekeeping mentor near you who is willing to allow you to shadow them.

I learned so much by apprenticing under a mentor and now I am mentoring others. It is an amazing feeling when you know what you know about beekeeping and the confidence that comes with it.

In the next chapter, you will learn minimal disturbance methods that will help you before, during and after hive inspections. Among them are ways to help in minimizing risk to your highly-valued queens as well.

CHAPTER 8

THE 7 METHODS OF MINIMAL DISTURBANCE

A few years ago I came across some interesting beekeeping concepts when I was considering keeping bees in a Warre style beehive.

I won't go into great detail about that particular hive style but it is a style that allows honey bees to have a more natural environment than the one that Langstroth hive styles offer. The primary difference is that while the Langstroth style beehive is inspectable, the Warre style is not.

Although I ultimately decided to not pursue using the Warre style beehive, what I learned from it has completely affected the way I operate and manage my Langstroth style equipment for my bees and I'm going to share that with you now. There are basically two types of beekeeping styles: Anthropocentric and Apicentric.

Anthropocentric is a beekeeping style that focuses on the convenience of the beekeeper first and the honey bee second. This means that the management style and the equipment are designed and utilized in such a way as to cater to the beekeeper's needs first and foremost. The prefix 'a*nthro-*' means *man* in Greek. '*Centric*' means centered. You probably got that huh?

On the other hand we have the **Apicentric** beekeeping style in which the well-being of the honey bee comes first and the convenience of the beekeeper second.

This style focuses on specific, intentional beekeeping practices that honor, respect and maintain the honey bee colony as the central focus in order to maintain a thriving, healthy and productive colony.

The prefix 'a*pi*-' means bee, as in *Apis Mellifera (the scientific name for the honey bee.)*

Because of my apprenticing years in the commercial beekeeping industry, I now realize that the system we used then was very Anthropocentric. We used harsh chemicals to treat for mites, fed them pesticide-laden high fructose corn syrup as *'food'*, disturbed the colonies on a frequent basis with work and did our best to rush and complete that work in order to get to the next apiary.

I came to the conclusion that honey bees were treated as a commodity in the commercial industry by many beekeepers but certainly not all of them, rather than as the spiritual, amazing power house animal that they truly are. As a disclaimer, I feel the need to point out that while my mentor was considered a commercial beekeeper, he has a great respect for his bees and I am proud to continue that essence in my operation.

Until I had considered using a Warre hive style however, I had never seen the distinction in the two beekeeping styles. Once I had made the distinction and decided which one I wanted to practice for myself with my own beehives, I decided to create something that I call *The 7 Methods of Minimal Disturbance* as a blog post for my website: BeeNativeHoney.com.

The idea behind these methods that I am about to share with you is simple: a honey bee colony will thrive more when there is minimal disturbance and that by using intentional, high quality methods that make a bigger difference rather than frequent disturbance using lower quality practices without aim or goal, keeping bees is much more rewarding as a result.

By disturbing a honey bee colony less frequently, it is allowing them to maintain what is called a *homeostatic environment*. A homeostatic environment is one in which there are many small parts, each with their own task, working together and contributing to the greater whole for the highest efficiency and the greater good of the colony in every case.

Every time a beekeeper does a hive inspection he or she disturbs that homeostatic environment which takes time for the colony to recover from. So here are seven methods in no particular order, meant to become a habitual practice for the mindful beekeeper at all times.

Method #1—Anticipate & Prepare

This is definitely one of the core principles that uphold intentional, mindful and successful beekeeping. Enough cannot be said about the beekeeper who is motivated and diligent enough to continually educate themselves, ever learning more, how best to prepare when implementing a plan.

Despite the intricacies of the plethora of beekeeping tasks, this method is fairly easy and straightforward, involving only two steps to apply to each goal or seasonal task as you work throughout the year:

Step #1—Anticipate. To anticipate a need or something that needs to be done means that you are taking an active role rather than a passive one when it comes to your duties as a beekeeper. More than that, it means that you are actively thinking and preparing for things that need to be done before they actually need to be done.

In the world of beekeeping, anticipation is your super power. Anticipation applies to many things including the following as a few examples:
- Seasonal tasks.
- Splitting bees requires preparing equipment ahead of time.
- Supering hives for a honey flow means needing to have honey supers ready to go before the flow.
- Harvesting honey means that you need your work area ready to extract the honey.
- Treating for mites means that you need to have all the proper things ready and be aware of the weather forecast.
- This list could go on and on about what tasks to anticipate for. Even just a routine hive inspection to simply check for a laying queen or the hive's stores means that you need to have smoker fuel, smoker, lighter, hive tool and protective gear ready. I hate to think of how many times I have shown up at one of my apiaries without a lighter. It irks me. I am irked.

Step #2—Prepare. This is the implementation part of this method, putting into action the plans you've carefully decided to work towards.

This step involves getting the proper tools ready that you will need to use. For example if you're treating for mites using an oxalic acid vaporizer, then you need to get that ready, making sure it's clean and that you have your oxalic acid crystals, battery or generator, respirator and gloves and of course the usual hive tool, smoker and protective gear.

What is especially unique is that we beekeepers are a special crowd in that we are inventive, creative and love to try new things to test our ideas against.

With this method I could give you examples all day long of things to anticipate and how to properly prepare for them but you are intelligent enough to break this down on your own if you take the time to do so. No matter what, you will definitely begin learning exponentially. But for the case of what this book stands for, I will give you some of the major things that this method is designed to Anticipate & Prepare for. Things like:

- **The weather.** Only certain weather conditions are ideal for working bees.
- **All four seasons**; each season must be anticipated during the season before it. (*Prepare for winter in the fall*).
- **Anticipate the growing season for bees** and the waning season, when their populations begin to naturally diminish because the queen is slowing down her egg-laying rate. This is primarily done through equipment management.
- **Honey flows.** To take full advantage of a honey flow, you have to have healthy hives with large populations of worker bees and you also need to have your honey supers ready to go before the flow, not later.
- **Winter time** is a huge factor to consider when preparing your bees ahead of time. Getting them ready in the late summer through fall is crucial to getting them through winter well.

Method #2—Rain or Shine, Do Things on Time

The primary message here is that beekeeping is not a procrastinators game. Honey bees work season by season according to weather and climactic conditions on a day by day scenario. They never put things off to prepare themselves for the future.

Nature also works according to the seasons, with blossoms blooming on time more or less year after year and the bees work in tandem with this cadence. As beekeepers we must do the same.

If we are to take advantage of what our amazing little bees produce for us humans as a species, then we cannot forego tasks that are in our job title. The biggest or most impactful task needing to be done on time is pest & disease management.

Pest and disease management is one of our most crucial roles as beekeepers. By keeping pests and diseases under control using preventative measures, we are giving our honey bees the best fighting chance that they have to be a healthy, thriving and productive colony.

By no means am I telling you to work in the pouring rain. I will be the first to admit that I am not going to do bee work in the rain. What I mean by *"Rain or Shine"* is that no matter what, you need to do your best to do the proper seasonal tasks as close to on time as possible. Do not dilly dally because people will call you a dilly dallier if you do.

One autumn a few years back, I hit a bout of depression and failed to treat for mites. I lost a lot of hives that winter with very few that survived until springtime. Depression is not the easiest thing to deal with but I swore to myself that I would never allow that to happen to my bees or to myself again.

Method #3—Observe & Learn

So much can be learned about honey bees simply by observing the hive entrance, by observing what bees are foraging from and by watching their behavior as a whole from season to season.

One of my very favorite books that teaches how to uphold this method is a book originally written in German but fortunately translated into English. It's called, *At the Hive Entrance* by H. Storch. The book is designed with two columns on each page. The left hand column lists the observation made with the possible explanation of the hive's condition in the right column.

The author even took it a step further to organize the book into seasons so that you can flip from spring through winter.

An example of this may be that the observer notices the bees coming back to the hive in strong forces with a grey/green colored pollen on their legs. The possible explanation, since this would be in the springtime section of the book is that maple is now blooming and the bees are taking full advantage of it. What this means for the beekeeper is that with the influx of pollen coming in that the queen is increasing her egg-laying rate, thereby increasing the colony population in the weeks to follow. This external observation tells the beekeeper to expect a large, growing population inside the hive with lots of activity.

I remember one time when I was observing one of my beehives in Florida. The bees were bringing in purple pollen. It was really quite beautiful and I couldn't help but wonder what plant they were gathering it from. This led me to learn about a plant I had yet to identify as a pollen source for bees, specifically a swamp plant called *Pickerel Weed*. Not only did I learn this new plant and where it grew but also what time of the year that I could expect it to begin blooming as a source of pollen for my bees each year to follow.

Method #4—Smoke the Hive, Help Them Thrive

A smoker is not just for the beekeeper, it's for the bees as well. It helps to keep the bees away from equipment contact areas where they might otherwise be smashed and unintentionally killed during a hive inspection.

Many people believe that a smoker either puts the bees to sleep or calms them down. If you thought your house was on fire would you calm down? Nope. Smoke neither calms them down nor puts them to sleep.

Smoke used in small, reasonable amounts does one very simple thing: it covers up the bees' alarm pheromone. Honey bees work by smell and when a beekeeper comes around, ripping off the roof of their house, they have good reason to sound the alarm and go to red alert.

Using a smoker in the entrance and a little under the hive cover before fully entering the hive will go a long way to make conditions easier for the beekeeper to work and for the bees to not go ballistic on your butt.

Like I mentioned earlier, the smoker is also a tool in that it is used to coral bees away from areas that you *do not* want them to be or to coral them into areas that you *do* want them to be. This is most useful after a hive inspection when you are putting everything back together. Smoking the bees away from the box edges will help to minimize the amount of bees that get squashed as you stack boxes back on top of each other and place the hive cover back on afterwards.

Method #5—Go In with a Plan

Curiosity is a high-risk reason for a hive inspection. You only have one queen; and if you accidentally roll her between frames while removing one or squish her between two frames while going through the box, then you have set the colony's productivity back for at least 4–6 weeks.

It is vital that you perform hive inspections with legit seasonal tasks and goals in mind rather than simply for curiosity's sake.

Your plan will also keep you in check. Let's say that you recently made new splits and it's been two weeks and now it's time to check for laying queens. The plan is simple: go into each split, check for *evidence* of a laying queen which would be seeing eggs down in the cells. If you happen to spot the queen then that's a bonus but once you have confirmed a laying queen then it's time to close up the hive and move on to the next one.

Do not be the type of beekeeper that thinks you have to locate the queen every single time you perform a hive inspection.

Method #6—Everything in Its Place

Never ever divide up the brood nest with undrawn frames or frames of solid honey. Never. Part of the colony's ability to maintain a homeostatic environment is by maintaining a consistent temperature in the brood nest in order for the developing larvae and pupae to fully develop into adult honey bees that will ultimately contribute to gathering resources for the colony.

For the most part honey bees are extremely good at structuring their resources. Brood has its place as does pollen and honey. Once you learn this structure at the various times throughout the seasons, because it does vary, you will know what order to leave the hive in. Of course, there are exceptions to restructuring a hive depending on very specific factors like having the bees draw out new frames during a honey flow but aside from splitting bees, where you take a few frames of brood from one hive to make another, you must never divide the brood nest.

Having said that, I must point out one exception. There's always an exception right? When you have a hive that is producing brood and honey like crazy, it's okay to take a frame of brood during a split and replacing it with a drawn out, empty frame of comb in order to give the queen more room to lay eggs in. However I would never do this with a weak hive.

Method #7—Don't Get Greedy

In my early commercial beekeeping days I saw beekeepers harvest every drop of honey from a hive just to replace it by feeding their colonies with pesticide-laden high fructose corn syrup. The reason was simple: you make more money for more honey.

It's not widely spread in the media today but one of the biggest reasons for the high mortality rates among honey bee colonies in the US is because of the commercial beekeeping practices.

If we and our bees are to survive into the next era of beekeeping successfully, it's going to be because we left the old ways behind and adopted sustainable beekeeping practices that work and actually create a sacred space of respect for our honey bee.

When it comes to adhering to this particular method, it's specifically aimed at any product you could possibly harvest from a beehive like honey, propolis, bee venom, pollen, royal jelly and beeswax.

It's up to you to decide whether to practice this method or not. It's also up to you to learn how to determine what a sustainable harvest amount looks like based on your particular area.

These are The 7 Methods of Minimal Disturbance. Based on sustainability, they help you to be a more mindful beekeeper, making each action count as you intentionally create your unique beekeeping style and operation whether you have two hives or two thousand. These are the qualities that help to develop your intuition as a better beekeeper.

So there you have it. I have found that just about anything else I could come up with falls under one of these methods. Remember, these are not something to do in a particular order but rather they are methods to practice and always keep in mind, developing them as good habits every day.

CHAPTER 9

Mindfulness & Intention Allows for Intuition

A part of beekeeping that is extremely overlooked, and something I have been guilty of overlooking as well, is being mindful and working your bees with *intention*.

For the majority of my years as a beekeeper, I have approached honey bees with fascination of course but mostly in a transactional way. Loading the truck with the tools and equipment that I need for my specific goals, I would set off to do my bee work for the day. Entering the apiary I simply just worked through my hives with haste and tried to get to the next apiary in time to do the same. That is, until this year (*2022*).

Suffering hive loss unlike any time before, I had to take a serious look at the way I do things. Spending months developing a new management system for my beekeeping operation, there was one thing that I was missing and I did not realize it until my fiancee at the time and now my wife, did something that I will never forget.

Having just established a new apiary with sixteen 5-frame nucs, I brought my fiancee Haleigh to my new location, after all, she was the one who made my new apiary possible through her family connections. As we entered into the bee yard, she refused to wear any protective gear and immediately started verbally expressing love towards my bees, using her hands to mimic spreading the love throughout the yard.

I immediately felt a mix of emotions; gratitude that I had such an amazing woman that would express her love towards *my* bees, yet also mixed with the feeling of *'why hadn't I thought of that?'* I mean after all, I have always appreciated my bees, being thankful for them, I felt that I had a greater respect for them than most peoples' attitudes about bees in general but to love them? It had never occurred to me.

Thus began my exploration of approaching my bees with purpose, mindfulness and intention. In my opinion, when anything you want to do in life is approached with an attitude of intention, greater results occur than would have otherwise.

How to Approach with Mindfulness & Intention

The way I now approach my bees is more than just walking up to them with good thoughts and feelings and a smoker in my hand. Here's what I do, broken down into 5 steps:

Step #1—Plan my goals for that day or week according to the seasonal tasks and gather all of the appropriate tools that I will need and load them onto my truck.

In order to do this, I have to plan ahead, slow down and think. I tend to get in a hurry and rush to work, inevitably forgetting something at home. This does not make for a very productive day and if affects my attitude and mood in a negative way.

Slowing down gives me the chance to become present with myself and seriously consider what my bees' needs are right now and couple those as best that I can with my goals. It is much more enjoyable than rushing off haphazardly forgetting the tools and equipment that I need.

The result is that I am better prepared in my mind, knowing what I am going to do and confident that I have all the tools with which to do it. By slowing down I get more done.

Step #2—Once I have arrived at my first apiary that day, I take the time to simply open my mind and heart to be present with myself and the bees, once again taking my time and *feeling* my reason for being there. A good way to practice this is by simply watching the hive entrances' activities before getting my tools out. It's like checking in with the bees first to see what they are doing and tapping into their energy level.

The most important part of this step is *feeling* and *allowing* gratitude to arise for my bees and everything that they do for me. This sets my heart in the right place to work my bees so that I remain in my intention.

Slow, deliberate movements go a long way in beekeeping. Once I have taken the time to be present, slow down and ground myself in my surroundings and what is going on during that specific time of the year, I begin the next step.

Step #3—Now that I feel confident that I am fully present in the moment with my bees, the environment and my goals for the day, it's time to get my tools out. This step always involves lighting a smoker and getting it going really well before doing anything else.

As far as the other tools, I have a hive tool in my back pocket and I also have my pest and disease management tools at the ready. Or I might be making splits or adding honey supers in which case I have the equipment ready in the back of my truck.

Step #4—In this step it's time to get to work. It's important to smoke the entrance to each hive and under each hive cover before doing my in-hive inspections, one hive at a time.

Now it's time to go in with my plans in mind which is one of *The 7 Methods of Minimal Disturbance* as you read about in Chapter 8. During in-hive inspections it is crucial to keep in mind two things:
- The well-being of your little bees and,
- Your goals for that day.

Keeping these things in mind will help to keep you focused as you take your time, giving the same care and attention to each hive as you methodically work through your colonies. Honey bees are tiny and fragile. They may have a stinger but they are quite docile most of the time and need to be handled like a package labeled "*Handle with care.*"

The results will be evident in the benefits for the bees, your accomplishments for the day and how you feel for having accomplished your goals. The way you feel is what the last step is all about.

Step #5—The feelings of accomplishment and having completed what you were there to do will give you some pretty good feelings. This is also known as pride. This is the good kind of pride. Feel it and let it be the feeling that you leave your bees with that day. This allows you to mentally keep up with your bees as the seasons progress, lessening the stress of feeling behind.

Leaving the yard slowly and deliberately, and projecting those good feelings of pride and gratitude towards your bees is a very high vibrational energy to leave your bees with.

This is intentional beekeeping that keeps in mind the well-being of your bees at the forefront of your operation no matter how big or small.

In my 21+ years of beekeeping, there has not been any other practice that has created the benefits that I want for my bees and myself as Mindfulness & Intention have.

Part of beekeeping is having the right and creativity to stumble your way into something that works for you. Well I have had plenty of opportunities to stumble but I prefer to focus on getting back up. Failure isn't the end, it's part of the process.

Results of Mindfulness & Intention

The primary reasons that most beekeepers have bees are for the products of the hive:
- Honey
- Propolis
- Pollen
- Bee venom
- Beeswax
- Royal jelly

These are all good things, each with their own benefits and price tag. However they should not be our primary reasons for keeping bees. Ideally speaking, all of these beneficial products of the hive should be a *result* of good beekeeping practices rather than the goal.

Results are evidence of whether something works well or does not work well. Practicing this over and over time and again will eventually build up your intuition about how to properly maintain a beehive in the best way possible.

Finding a method that works for you, for your bees and in your specific area is a unique practice that does not necessarily work for someone else in a different location. While there are certain general guidelines to successful beekeeping, there are slight variations as well which is one more thing that makes this trade more unique than any other.

Essentially you are becoming your own coach, your own mentor, and your own source of knowledge as you gain experience using your intention.

Doing Nothing Is Sometimes Best

In a fast-paced and ever-demanding world, the concept of doing nothing may seem counterintuitive. We are often conditioned to believe that taking immediate action is the key to solving problems. However, there are situations where doing nothing can be the wisest course of action. Sometimes, inaction provides us with the clarity, perspective, and space needed to find the most effective solution.

Firstly, doing nothing allows for a pause, giving us time to reflect and evaluate the situation. When faced with a complex problem, our initial reactions may be fueled by emotions, stress, or impulsiveness. By stepping back and refraining from immediate action, we gain an opportunity to detach ourselves from these emotions. This mental distance enables us to approach the problem from a more objective standpoint, enabling clearer thinking and better decision-making.

Additionally, doing nothing can be a strategic move in situations where intervention may exacerbate the issue. Certain problems have a tendency to resolve themselves naturally if given time and space.

Doing nothing can also foster innovation and creativity. When faced with a challenging problem, our minds naturally seek solutions. By allowing ourselves moments of stillness, we create space for ideas to incubate, grow and new perspectives will begin to emerge. Often, the best solutions come when we least expect them, during moments of relaxation or mind-wandering. Doing nothing provides an opportunity for our subconscious minds to process information, make connections, and generate fresh insights. There are times when you may not know what to do as mentioned previously in this book. This is okay and simply another part of beekeeping.

This is not necessarily a time of doing nothing per se. Take this time to *refrain* from taking immediate in-hive action and manipulation and use this time to sit, observe, listen and be open to a solution coming *to you* instead of trying to come up with one yourself. Try being still and try to keep a clear mind . One way that I like to do this is by sitting near by, watching my bees come and go and simply listen to the buzz in the air. It's a way to get lost and free from your thoughts, especially if you're an over thinker.

Patience is the next part of this equation. The answer may not come to you immediately or even that very day. It may take a good night's sleep to slow the momentum of the problem so that you can find yourself with an open mind to receive the solution and often it presents itself to you when you are no longer thinking about it.

I believe that it's safe to say that we have all experienced these epiphanies when we least expected them. When the solution does show itself to you, it may be time to plan on how to take action, followed by implementing the solution just as we learned in chapters 2 & 3.

While it may seem counterintuitive, doing nothing can be the best solution to a problem in certain circumstances. Taking a step back, gaining perspective, avoiding unnecessary intervention, conserving resources, and fostering creativity are all potential benefits of embracing inaction. By recognizing the value of doing nothing, we open ourselves up to the possibility of finding more effective and efficient solutions to the challenges we face and we get more accomplished as a result.

In conclusion, the role of intuition in beekeeping is of great importance, intertwining experience, observation, and instinct into a powerful tool for successful apiary management. Beekeepers who possess a keen sense of intuition can tap into a deeper understanding of their colonies, enabling them to make informed decisions and navigate the intricate dynamics of a beehive with confidence.

Intuition allows beekeepers to sense subtle cues and interpret the behavior of their bees, providing valuable insights into the health and well-being of the colony as easy as reading a book cover. This intuitive understanding helps in identifying potential issues or abnormalities, such as issues with disease or changes in foraging patterns, allowing appropriate intervention and prevention of further complications.

Even more, intuition empowers beekeepers to anticipate changes in environmental conditions and adapt their management practices accordingly. I know that I am also observing and adapting to my observations with incredibly positive results most of the time. By observing natural phenomena, like blooming patterns, weather fluctuations from year to year, intuitive beekeepers can make informed decisions on hive placement, timing of inspections, or when to implement supplementary feeding.

Intuition also plays a vital role in decision-making during hive inspections and honey harvesting. Experienced beekeepers develop a sixth sense that guides them in gauging the strength and health of a colony, determining the appropriate timing for expansion or honey extraction. This intuitive assessment helps prevent disruption to the delicate balance within the hive and ensures the well-being of the bees.

However, intuition alone is not sufficient. It should be complemented by best practices, and ongoing learning. By combining intuition with experiential-based beekeeping techniques, beekeepers can achieve a harmonious synergy, allowing them to optimize colony management while respecting the inherent nature of the bees.

In the world of beekeeping, intuition is a valuable asset that enables beekeepers to connect with their colonies on a profound level. It is a skill that is honed over time, drawing on experience, observation, and an innate understanding of the bees. Harnessing the power of intuition, beekeepers can navigate the intricate world of honeybees with greater confidence, fostering healthy and thriving colonies for the benefit of both the bees and the beekeepers themselves.

Mindfulness & Intention Allows for Intuition

Stop.
Look.
Observe.
Listen.
Think.
Ponder.
Be still.
Receive.
Take action.
Repeat.
This is beekeeping.
Have fun!

Jonathan Adam Hargus
—Small scale, Sustainable commercial beekeeper

References

American Honey Plants, (1978), Frank C. Pellett, Dadant & Sons, Inc, USA.

At the Hive Entrance, (2019), H. Storch, Unknown Publisher, USA.

The Spiritual Foundations of Beekeeping, (2017), Iwer Thor Lorenzen, Temple Lodge Publishing, Essex, England.

Glossary of Beekeeping Terms & Lingo

Apiary—The place where you keep and maintain your beehives.

Apiculture—The art of beekeeping that goes beyond standard practices into the world of maintaining honey bee colonies in a sustainable manner.

Apis Mellifera—The scientific name for the Honey Bee.

Artificial Insemination—The act of controlling honey bee genetics by artificially inserting specifically chosen semen into a queen.

Beekeeper—One who educates him or herself, seeking better ways of keeping beehives healthy and thriving.

Bee Dance—Physical patterns of movement communicated by field foragers to other foragers, indicating precise locations of nectar yields.

Bee Space—The necessary width inside of a hive that bees move to and fro through in order to facilitate colony work. A bee space is approximately ⅜".

Beeswax—A substance produced by honey bees as the building blocks and foundation of the colony. Beeswax is formed into comb, where food is stored and baby bees are developed.

Boardman Feeder—A feeder designed to use a jar to feed colonies through their front entrance.

Brood Chamber—The box used to maintain the area where the queen lays eggs and the nurse bees raise her young into fully matured adult bees.

Brood Nest—The physical area where frames of brood are located.

Burr Comb—Comb built by honey bees that fill in gaps, having no discernible form or function.

Capped Brood—The third and final stage of brood when older larvae develop into pupae.

Capped Honey—Honey that has been ripened to a point of low moisture content and capped over as a way to store the honey as food for later use.

Castes—The three castes in a honey bee colony are Queen, Worker and Drone.

Cell Cup—Also called "buttons," cell cups are small cups of wax that are kept in case the colony needs to raise a queen for the purpose of supersedure or swarming.

Cell—Each individual cell on a frame is called a cell.

Chilled Brood—When brood gets too cold and dies it is considered to be chilled.

Cleansing Flight—Bees are extremely hygienic and will take cleansing flights outside of the colony to use the bathroom.

Colony—The name for the entire entity of a bee hive; workers, drones and queen.

Dearth—A nectar dearth occurs when very little is blooming at specific times, usually resulting in robbing.

Drawing Comb—The act of building beeswax onto frames, generally done when sufficient resources are available through bloom.

Drone Bee—The boys. They exist purely for mating with virgin queens.

Drone Comb—Cells that are larger in diameter than worker cells, where drone eggs are laid by the queen.

Fanning—There are many reasons why honey bees fan their wings but in each case they are either controlling temperature in the hive or spreading pheromone for the purpose of communicating to the entire colony.

Festoon—The cool word for when honey bees hold hands, hanging from one another and the frames during the process of drawing beeswax and building it out.

Foundation—Whether it's plastic or beeswax, foundation is the beginning of what honey bees draw wax out from.

Frame—Frames are what each type of bee box holds for the purpose of housing and maintaining a honey bee colony.

Hive Stand—The base on which a beehive is placed upon for level stability.

Hive Tool—A special pry bar made specifically for prying apart hive boxes and frames during hive inspections.

Hive—Technically speaking, a hive consists of the materials in which a honey bee colony lives.

Honey Bound—During a honey flow there is a lot of incoming nectar. A colony can become honey bound when they have no more room to store honey and are full up.

Honey Flow—A honey flow occurs at a time of year when there is so much blooming from major sources of nectar, that the honey bees make more honey than they need.

Inner Cover—When using a telescoping cover, it is necessary to employ an inner cover first in order to create ventilation flow.

Langstroth Hive Style—Considered the standard hive style in the US, the Langstroth style hive was the first created as "inspectable" and pretty much redefined beekeeping.

Larva—After a queen's egg hatches it becomes a larva.

Laying Worker—Under certain circumstances that are less than ideal, worker bees can and will begin to lay eggs but they always result in drones because they are unfertilized.

Mating Flight—Once a queen has newly emerged as a fully-developed adult, she must go on a mating flight where she will mate with 10–20 drones, giving her enough sperm to lay eggs for the remainder of her days.

Nectar—The attractive liquid offered by plants via their blossoms that honey bee gather and turn into honey.

Nucleus (Nuc)—This is considered the core base of a honey bee colony, usually consisting of frames of brood, frames of food and a laying queen and of course thousands of worker bees and some drones.

Nurse Bees—Young worker bees whose duty it is to tend to the developing brood, ensuring their food supply and warmth.

Orientation Flight—Once a honey bee is old enough they will take an orientation flight by leaving the hive and orienting themselves visually on their home location.

Package Bees—There are several ways to obtain honey bees and 3-pound packages of honey bees is one of those. A package of bees consists of a caged mated queen and approximately three pounds of worker bees, surrounding a can of feed for transit until they reach their final destination.

Pollen—Pollen is something that flowers produce and that pollinators like honey bees pass from blossom to blossom, as well as use it as food.

Propolis—This is a resinous substance collected from plants that bees collect and use to "glue" things together and also to maintain a sanitized home.

Queen Bee—The matriarch of a colony, the queen is responsible for laying eggs and maintaining the colony's population.

Queen Cell—A queen cell is when there is a live larva or pupa within a special cell that will ultimately grow into and become a queen bee.

Queen Excluder—Excluders are used to keep the queen in the brood chamber and out of the honey supers so that she does not lay eggs in the honey supers intended for harvest by the beekeeper.

Queen-less—The condition when a hive is without a queen.

Queenright—The condition when a hive has a laying queen.

Ripening—The process of lowering the moisture content in nectar between 15–18% for the purpose of creating honey.

Robbing—When certain environmental conditions are met such as a dearth, honey bees and other would be predators will seek honey from wherever they can find it, even if that means robbing it from other honey bee colonies.

Scout Bees—Worker bees that seek out potential homes for when their colony swarms from one location to another.

Smoker—A tool vital to successful hive inspections, smokers are a way to cover the alarm pheromone of honey bees.

Super—The short word for Supernumerary, something that is temporarily placed onto something else, in this case a honey super is used during honey flows and are otherwise stored when not in use.

Varroa Mite—The scourge of beekeepers, Varroa mites are tiny parasites that feed on the brood of honey bees, thereby spreading viruses and wreaking havoc.

Washboarding—The action of many bees on the outside of the hive, moving back and forth using their mandibles to do something that I don't quite understand. Any hive that is seen washboarding can be considered quite healthy and thriving. Weak hives do not washboard.

Winter Cluster—The dormant state of honey bees, during the winter, bees tighten up into a ball of bees to maintain a warm, dry and comfortable condition for survival through the cold months.

Worker Bee—All workers are females. Worker bees perform all the tasks necessary to maintain colony efficiency for the good of the colony.